NEIL GOULDING

JET MECHANICS
& HYDRAULIC
STRUCTURES

Theory, analysis and design

A collection of illustrated works providing a manuual on jets and streams and the design of hydraulic structures for conveyance, control and distribution of water in irrigation, drainage, hydropower, water supply and other projects worldwide.

Front Cover:
Wimbleball Dam and Reservoir
in Devon
(South West Water)

Back Cover:
Dam Spillway and Gates,
Curhuaquero Project, Peru.
(Balfour Beatty)

Professor S K Al Naib
BSc (Eng) PhD (Lond) ACGI DIC CEng FICE FIMGT
Professor of Civil Engineering and Head of Department
University of East London
Great Britain

BOOKS FOR WORLDWIDE APPLICATION

HYDRAULIC BOOK SERIES

Self-learning with user friendly books is a great source of satisfaction, and will add value to your personal development. The hydraulic book series shows you how to master all the basic skills and develop them imaginatively. Practical step-by-step procedures and inspirational diagrams combine to give you all the information you need to create a successful path in your studies.

Book 1: Fluid Mechanics, Hydraulics and Environmental Engineering.

Book 2: Applied Hydraulics, Hydrology and Environmental Engineering.

Book 3: Jet Mechanics and Hydraulic Structures.

Book 4: Experimental Fluid Mechanics and Hydraulic Modelling.

The environmental chapters include surface and groundwater, water power, water supply, public health, irrigation and coastal engineering.

HERE IS A GREAT DEAL FOR YOU

- Ideal for student-centred learning.

- Essential information for students, lecturers, technical staff and practising engineers.

- A unique and comprehensive set of books.

- Have easy to read formats.

- Contain excellent illustrations and diagrams.

- All books at attractive prices.

Readers come from the universities, consultants, government departments, local authorities, water companies, etc. Whatever your use of these books, you will not want to miss out on the benefits of a set in your library.

See information on the UEL web site at http://www.uel.ac.uk or telephone 0181 590 7000 extension 2478/2531, or fax 0181 849 3423.

Please order through : Research Books, P.O. Box 82, Romford, Essex RM6 5BY, Great Britain

Review of Books by Dr John P. Grubert
BSc (Eng), DipIng(Delft), MPhil, PhD, MSc(Comp), EurIng, CEng, CMath, MICE, FASCE, FIMA

These books by Professor Naib, because of their user friendly style of theory, worked examples and exercises intermixed with high quality diagrams, fulfil a present day need by faculty lecturing in the subject areas of fluid mechanics, hydraulic and environmental engineering. They have been designed not only to encourage more classroom interaction between professor and students, but also to interest and enthuse students from diverse backgrounds into self-learning. Books 1 and 2 are unique because of the wide range of topics covered, the large number of worked examples and their presentation quality, which are not found in any other similarly priced books. I personally have read through each book in considerable detail, and with over twenty-five years lecturing experience, eight of those years as a full-time professor in US universities, can thoroughly recommend them to both instructors and students.

ISBN 1 874536 090

First Printing January 1992

Second Edition June 1998

Internationally Acknowledged Books by the Author

"Fluid Mechanics, Hydraulics and Envir. Eng"	ISBN 1 8745 36 066
"Applied Hydraulics, Hydrology and Envir. Eng"	ISBN 1 8745 36 058
"Jet Mechanics and Hydraulic Structures"	ISBN 0 9019 87 832
"Experimental Fluid Mech. and Hyd. Modelling"	ISBN 1 8745 36 090
"London Dockland Guide" Heritage Panorama	ISBN 1 8745 36 031
"London Illustrated" History, Current & Future	ISBN 1 8745 36 015
"Discover London Docklands" A to Z Guide	ISBN 1 8745 36 007
"London Docklands" Past, Present and Future	ISBN 1 8745 36 023
"European Docklands" Past, Present and Future	ISBN 0 9019 87 824
"Dockland" Historical Survey	ISBN 0 9089 87 800

The author is Professor of Civil Engineering and Head of Department at the University of East London, Longbridge Road, Dagenham, Essex RM8 2AS, Great Britain.

Preface

This book places on record a collection of research works by the author during the past quarter of a century on the behaviour of water jets and streams and various aspects of design and performance of hydraulic structures.

Over twenty selected papers with introductory chapters discuss a wide range of topics investigated from 1965 to today. Special attention is given to the application of the research to the design of engineering works. Guidelines have been presented for the design of structures for the conveyance, control and distribution of water. The essays constitute a first attempt to examine the flow of water jets with free surface and confined boundaries. More than 200 diagrams and illustrations depict the modes and details of various types of streams and equations which describe their laws and similarity to other types of free air jets.

It is hoped that the effort in compiling this book will make available for engineers who may be engaged in engineering design or research some information on the mysteries of fast running water.

It is an appropriate time to feel that the research has been set down for the benefit of future generations. They can look back and understand how we got to the position in which we found ourselves in 1990.

The book provides an authorative and unique reference for lecturers, undergraduates and postgraduate MSc students, research and practising engineers. It is suitable for use by consultants and other organisations engaged on the design, construction and maintenance of projects on irrigation, land drainage, hydropower, water supply and other water schemes in all parts of the world. The design procedures should form an enhancement of existing empirical techniques coupled with model testing.

No account of research and development would be complete without recording the fact that the contributions to knowledge have only been made possible as the result of considerable effort by many previous engineers and scientists who conducted valuable research in this field and contributed in varied ways to many facets of water and other fluid flow in nature.

An Invitation to the Reader

If you are a reader with unpublished research information on water jets and streams or relevant hydraulic structures, please contact:

Professor S K Al Naib
Head of Department of Civil Engineering
University of East London
Longbridge Road
Dagenham
Essex RM8 2AS
United Kingdom.

Telephone: 0181 849 3580
Fax: 0181 8493423

CONTENTS

PREFACE

PART I

JET MECHANICS
FREE TURBULENT JETS

Jet Boundary
Plane Jet
Round Jet

FREE TURBULENT JETS

Introduction

By a jet is meant a stream of fluid which travels far downstream of its source of supply with a velocity much higher than that of its surroundings. Such a jet may be produced by making a hole in the side of a tank containing water. Other liquid jets discharging into the atmosphere include jets from fountains, and fire hose nozzles; their behaviour has been studied by engineers and scientists for many centuries.

Air jets discharging into the atmosphere and water jets entering a large lake are however the sort of jets often engineers encounter in industrial design. There are, of course, innumerable boundary conditions. that introduce complexities. One such complexity is that produced by a solid boundary placed near the jet that interferes with the inflow of fluid required by the mixing process.

In civil engineering work, erosion of the bed and sides of channels and waterways is the outcome of many factors, but the principal cause is due to the fast jets and streams which are discharged downstream of hydraulic structures. Research has therefore been carried out to study the behaviour of such jets and to collect information in order to develop practical guidelines for engineers on ways of protecting banks of rivers, canals, irrigation and drainage channels against erosion. The research has covered water jets downstream of overflow weirs, sluice gates, pipe and tunnel outlets, channel expansions, etc. The results are discussed in other chapters of the book.

In this chapter a brief discussion is given of turbulent jets with no restricting walls, ie. the submerged free jets. These are of three kinds : jet boundary, plane jet and round jet. Experimental results are presented for the estimation of the increase in width and the decrease in velocity of these jets.

Principles of Jet Analysis

In order to analyse the flow pattern and the velocity distribution, geometrical similarity of the motion in the successive sections of the jet is assumed. Fig. 1 shows two velocity distributions at distances x_1 and x_2 from the origin of the jet. Similarity between the different distributions is said to be satisfied if a set of parallel straight lines in any direction and having their ends on the boundaries of one distribution is transformed to the next distribution still as a parallel set of straight lines.

It follows that if U_1 and U_2 are the central velocities at two successive sections, u_1 and u_2 are two velocities at distances y_1 and y_2 from the axis of the jet so chosen that

$$\frac{y_1}{b_1} = \frac{y_2}{b_2}$$

Then

$$\frac{u_1}{U_1} = \frac{u_2}{U_2}$$

This equation implies that generally u/U is the same function of y/b for all sections, or

$$\frac{u}{U} = f\left(\frac{y}{b}\right)$$

To relate the different variables in the problem, the opening for the jet is assumed of infinitesimal dimensions : a punctiform hole or a rectangular opening of infinitesimal width. This assumption together with that of the geometrical similarity lead to the correlation that b/x is constant, or

$$b = \text{constant x distance} = c\,x$$

This equation means that the width of the mixing zone is proportional to the distance from the origin of the jet. It applies to all types of free jets.

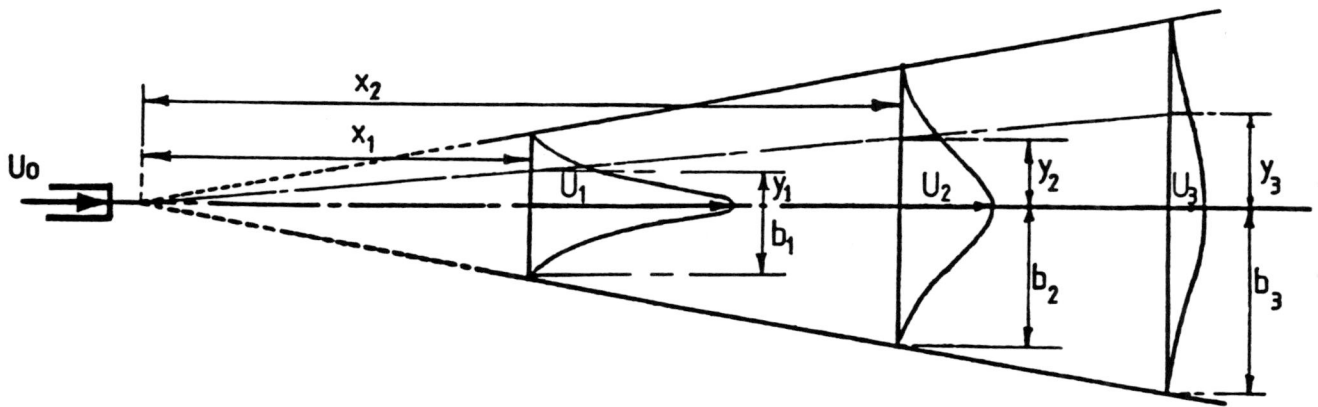

Fig 1 - Similarity of Velocity profiles in a free jet

According to the above assumption the velocity function can be written as

$$u = A \, f \left(\frac{y}{x} \right)$$

where A is an arbitrary constant whose value is to be determined according to the boundary conditions. The object of the mathematical solutions is to find in the different cases of flow the form of the above function.

Conditions within the jet were first investigated by W Tollmien on the assumptions that:

a)   The static pressure throughout the open jet is constant, and no vacuum exists;

b)   the viscous forces are small compared with the turbulent or inertia forces;

c)   the jet and the surrounding medium are of the same density;

d)   the jet is being discharged from a point source.

These assumptions are all closely approximated in many practical cases. With the exception of a numerical coefficient which had to be evaluated experimentally, Prof. Tollmien was able to derive a stream function with good approximation the actual pattern of the mean flow in each region, found by many experiments.

Regions of Flow

With reference to Fig.2., it will be seen that an initial zone of flow establishment exist beyond the afflux section of either a plane or round jet. Since the fluid discharged from the boundary opening may be assumed of relatively constant velocity, at the afflux section there is pronounced velocity discontinuity between the jet and the surrounding fluid. The eddies generated in this region of high shear result in a lateral mixing process which progresses both

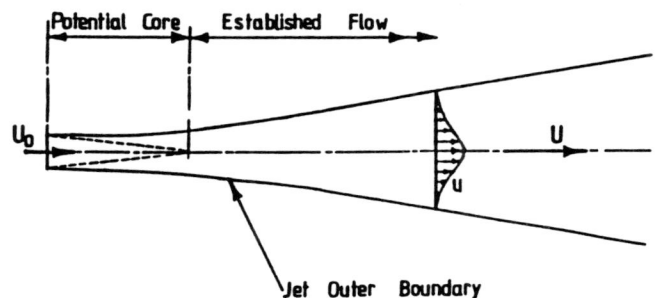

Fig 2 - Regions of flow in a jet

inward and outward with distance from the jet outlet.

Such lateral mixing produces a balanced action and reaction: On the one hand, the fluid within the jet is gradually decelerated; on the other hand, fluid from the surrounding region is gradually accelerated or entrained. As a result, the constant velocity core of the jet will steadily decrease in lateral extent, whereas both the rate of flow and the overall breadth of the jet steadily increase in magnitude with distance. The limit of this initial zone of flow establishment is reached when the mixing region has penetrated to the centre line of the jet. This zone of the flow is known as the potential core and normally ranges from 2 to 6 diameters from the outlet face.

Once the entire central part of the jet has become turbulent, the flow may be considered as fully established, for the diffusion process continues thereafter without essential change in character. Further entrainment of the surrounding fluid by the expanding eddy region is now

3

balanced inertially by a continuous reduction in the velocity of the entire central region. Such variation approaches a limit as the centre line velocity becomes of negligible magnitude a long distance from the original afflux section, the later extent of the eddy region - and hence the quantity of fluid within the jet - then being wholly out of proportion to conditions at the initial section.

Jet Boundary

A jet boundary occurs when a parallel stream of fluid meets fluid relatively of infinite extent and is at rest. The surface of discontinuity in the velocity of flow is unstable and gives rise to a zone of turbulent mixing downstream of the point where the jet is discharged. The width of this mixing region increases in a downstream direction, Fig.3.

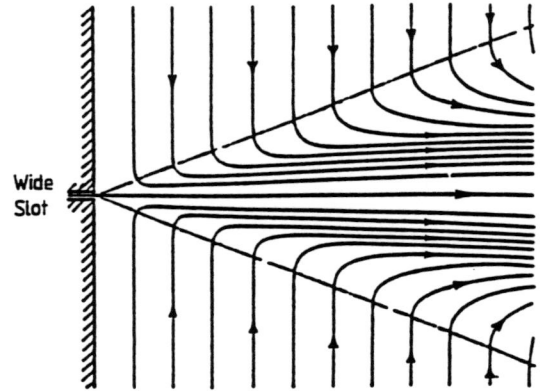

Fig 4 - Two-dimensional plane jet

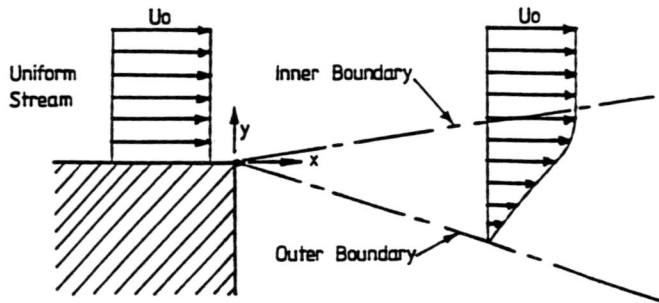

Fig 3 - Jet Boundary

Based on theoretical analysis and experimental measurements, the inclinations of the streamlines forming the limits of the jets are:-

a)   Boundary on the uniform stream side has an angle of 4 degrees 48 minutes which is equivalent to 1:12.

b)   Boundary on the atmosphere or fluid at rest side has an inclination of 9 degrees 48 minutes which is equivalent to 1:5.8.

c)   The streamline emanating from x = o defining the stream function ψ = O has an angle of 54.1 minutes which is equivalent to 1:62.5.

Two-dimensional Plane Jet

A plane jet occurs when a fluid is discharged from a narrow slit or opening and mixes with a medium at rest. (Fig.4) The results for an air jet from a two-dimensional long slot are given in Fig.5. The linear variation extends

up to 60d. The central velocity at a distance x from the plane of the opening is given by:-

$$3.64 \ U_o \sqrt{\frac{B}{X}}$$

where B is half width of the slit. The discharge passing down any section at distance x from the opening is given by

$$Q = 0.80 \ U_o \sqrt{\frac{B}{x}}$$

U_o is the original velocity, ie, the velocity with which the jet issues from the opening.

For the flow in the potential core the jet boundary results given above apply. The potential core replacing the uniform stream as shown in Fig.6. It can be considered as a uniform stream of finite surrounding fluid at rest on both sides. If 2B is the width of the opening, then the potential core ends at a distance L from the plane of issue as given by:

$$L = 12B$$

The width of spreading of the stream at a distance x from the plane of issue to the end of the potential core is given by:

$$b = 0.255 \ x$$

4

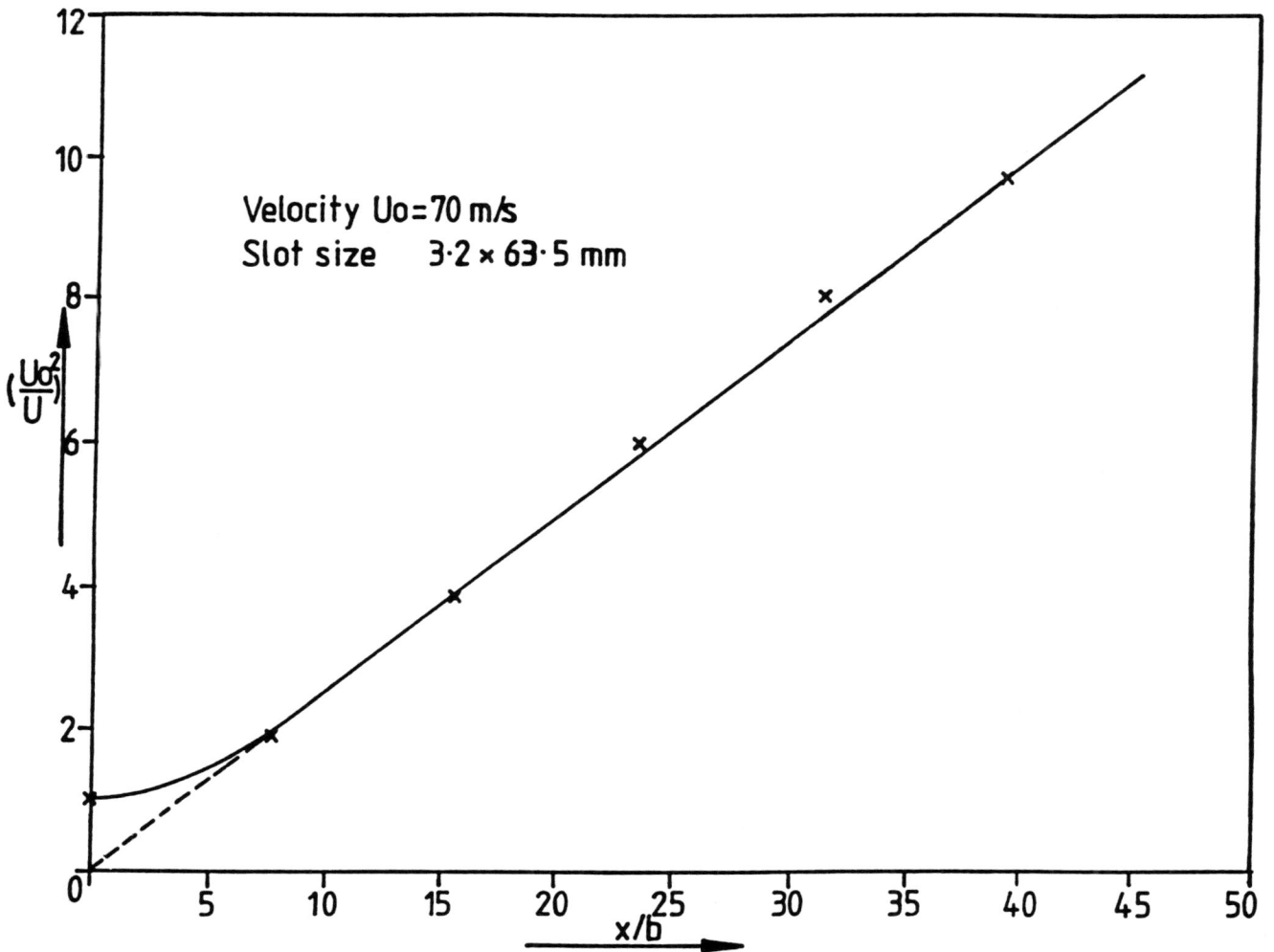

Fig 5 - Axial Velocity distribution in a plane jet

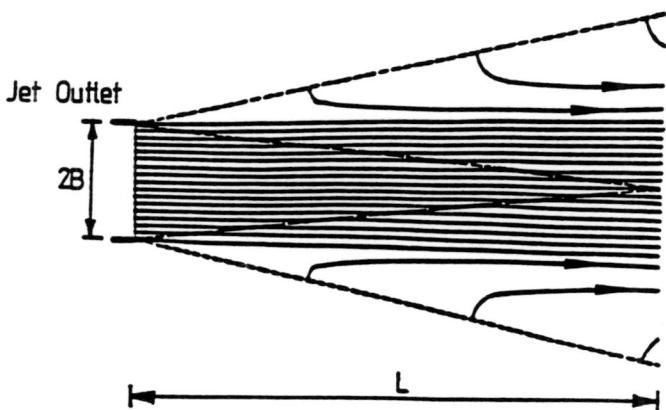

Fig 6 - Potential Core of a plane jet

which is equivalent to an inclination of 14.3°.

Three-dimensional Round Jet

When a stream emerges from a nozzle or a circular hole, the issuing cylindrical jet is symmetrical about the central axis and is termed a round jet, Fig.7. The flow in the potential core was analysed by Kuethe by a method of successive approximation who deduced the length of the core to be about 5 diameters.

Fig.8 shows a non-dimensional plot of the velocity ratio U/U_o against x/d for a circular orifice. In the graph the central velocity remains constant over a certain distance followed by a transitional length which joins to a straight line further downstream. The length of the

5

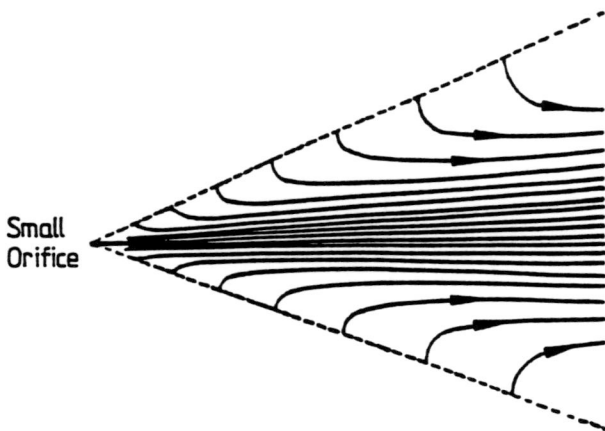

Fig 7 - Three-dimensional round jet

potential core as given by the horizontal part of the graph, is seen to be about 5 diameters in accordance with theoretical value predicted. The length of the transitional part of the curve is found to vary slightly such that the linear distribution begins at a distance ranging from 9 to 12 diameters. It appears that the higher the Reynolds Number of the jet, $U_o d/\nu$, the greater the mixing process and hence the shorter length of the transitional flow. The subsequent linear relation holds for distances exceeding 50 diameters. This variation is consistent with the theoretical relation, that the maximum central velocity varies inversely with distance from the origin.

The central velocity at a distance x from the plane of the opening is given by:

$$\frac{U_o}{U} = 0.152 \frac{x}{d}$$

where d is the diameter of the orifice.

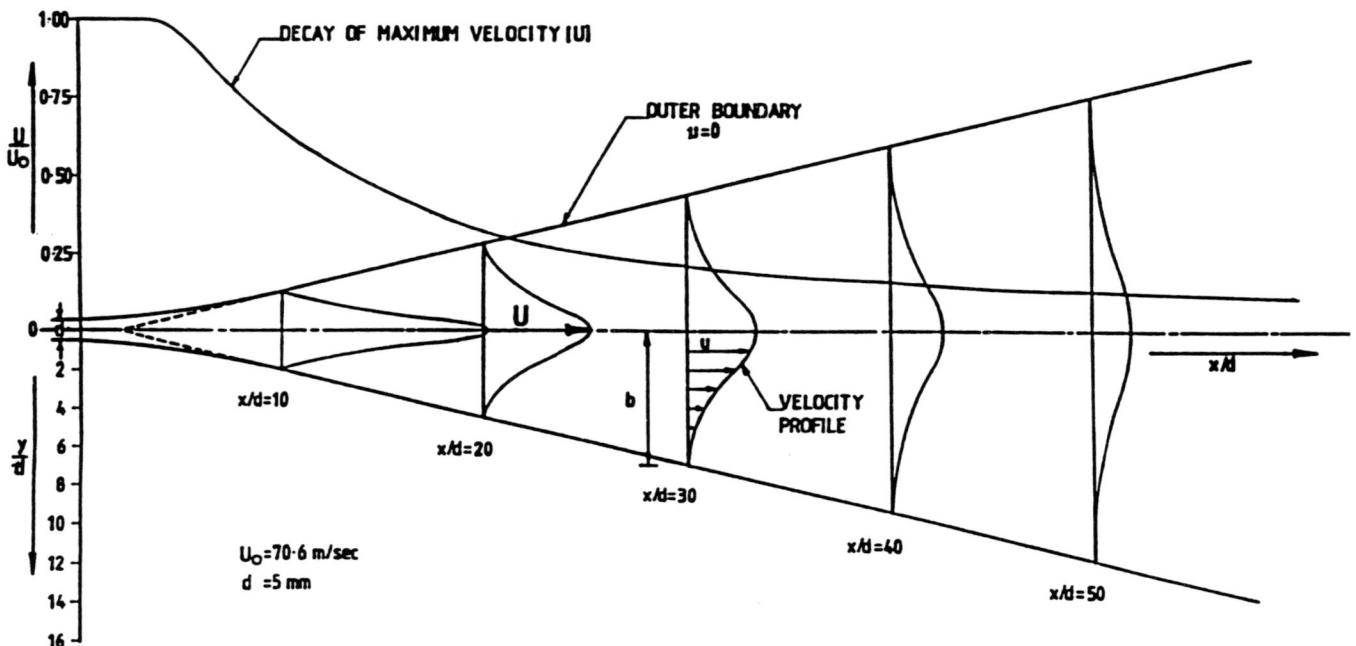

Fig 8 - Axial Velocity distribution in a round jet

6

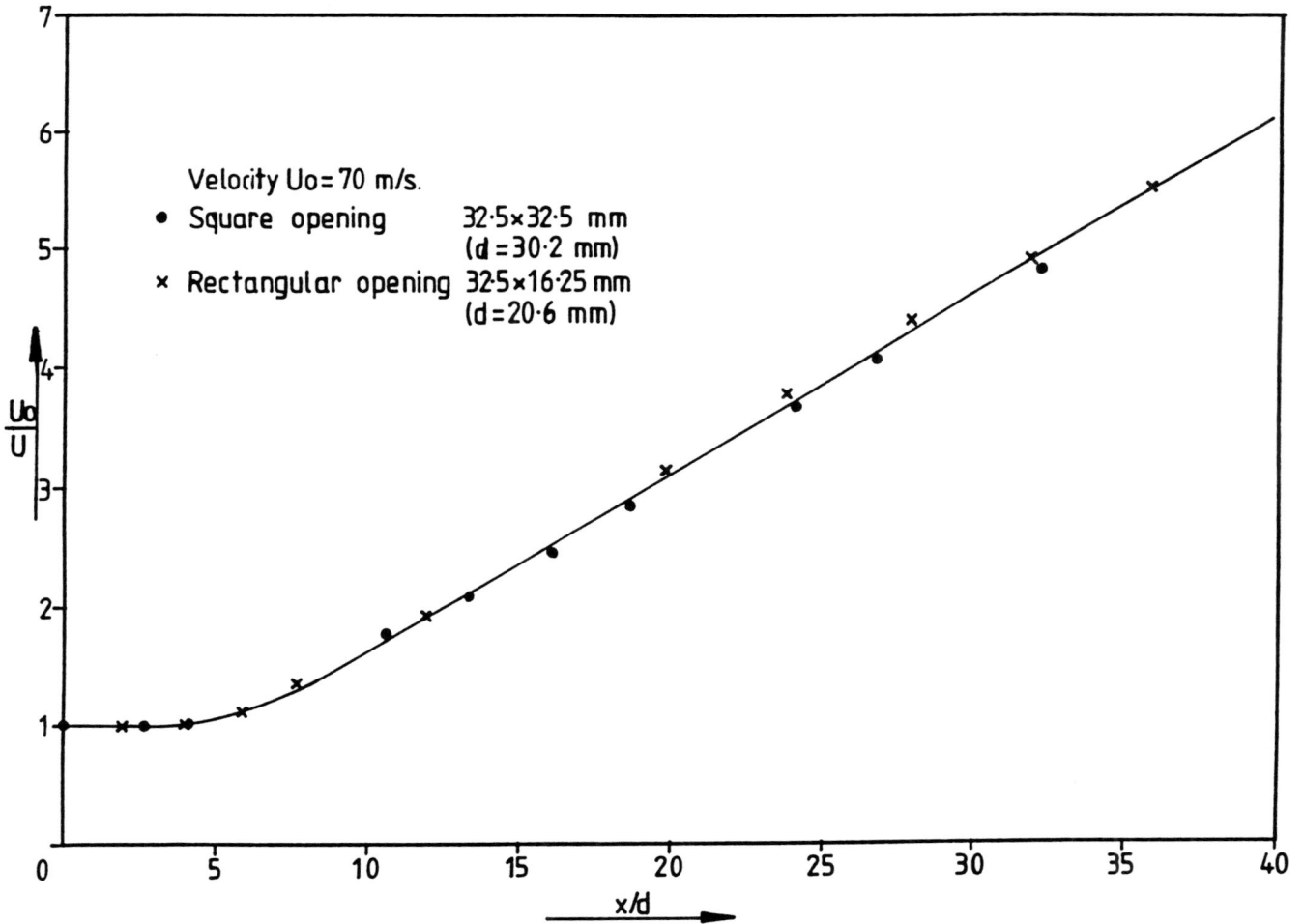

Fig 9 - Axial Velocity distributions in square and rectangular jets

Square and Rectangular Jets

The results for a square orifice and a 2:1 rectangular orifice are shown in Fig.9. For these openings the length of the potential core and the subsequent variation of velocity are practically the same as for the circular opening provided the equivalent diameter d is used. Beyond a certain width to depth ratio, approximately 5:1 the transitional length progessively increases for rectangular jets. It appears that the linear variation begins at the section where the width of the turbulent mixing is the same in all directions.

PART II

HYDRAULIC STRUCTURES

Channel Control Structures
Channel Transition Structures
Dam Spillway Structures

Channel Control Structures

Introduction

Flow controls and directional changes are normally needed in all stages of water conveyance and distribution from the water source to the final destination. These are achieved by the combination of constructing hydraulic structures in the system and organizing the operation of the system to ensure an adequate water supply. For a control structure, the elevation of the water surface can be predicted at a particular cross section for the given rate of discharge. Weirs, sluices and spillways may be cited as examples of such controls. Invariably, these structures induce changes in flow conditions from subcritical upstream to supercritical downstream. Fast jets and streams are formed which travel far distances before they are dissipated. Hydraulic energy dissipators are therefore incorporated in the design of such structures.

There are various methods for controlling the conveyance of water, namely upstream control, downstream control and the combination of both of these methods. For upstream control, hydraulic structures are designed to maintain a constant water level at the upstream end; the discharge capacity is controlled by fixed weirs or regulators with vertical or radial gates. The volume and level of the water in the upstream channel remains constant whatever the off takes along the distributary. In the downstream control system, the effect of the amount drawn is transmitted step by step upstream to the main supply, to be adjusted by automatic control structures according to the amount of water drawn downstream.

Drop Structures

In hydraulic schemes it is often necessary to provide control and drop structures for the following purposes:

a)   To flattern hydraulic gradients to canals which would otherwise be too steep because the average ground slopes are steeper than the allowable water surface slopes.

b)   To obtain command at specific points where the water surface has to be raised to enable offtaking main or branch canals to be supplied.

In case a) no regulation will be provided unless main or branch canals take off immediately upstream of the falls whilst in case b) regulation will always be provided.

Design Features

A drop structure is basically a weir (Fig.1), whose function is to maintain a certain required fall in canal or river level through virtually the whole range of operating discharges. In the case of a regulated fall it should also be able to hold up water to such a level that command upstream is obtained. In both cases, an abrupt fixed or varying drop in water level is artificially created and this fall in water level results in the conversion of potential head to kinetic head. This head must be destroyed within the solid boundaries of the structure or it will destroy the channel downstream of the structure by scour or erosion. If the channel is destroyed, the structure is likely to fail also, either structurally or as a result of undermining.

The weir is also a dam and is subject to overturning, sliding and uplift forces. It must be stable against these under all conditions. Most canal fall structures are low and if built on permeable foundations require sustantial length. There is usually no danger of sliding or overturning and it is not generally considered in design. Uplift is important, however, for unless proper provision is made the structure, or part of it will fail. Provision must also be made to prevent failure by piping, i.e. progressive washout of the soil beneath the structure.

A basic shape can evolve to standardize the procedure of calculations and to reduce the necessary hydraulic research to aspects of one model. The fall may consist of an

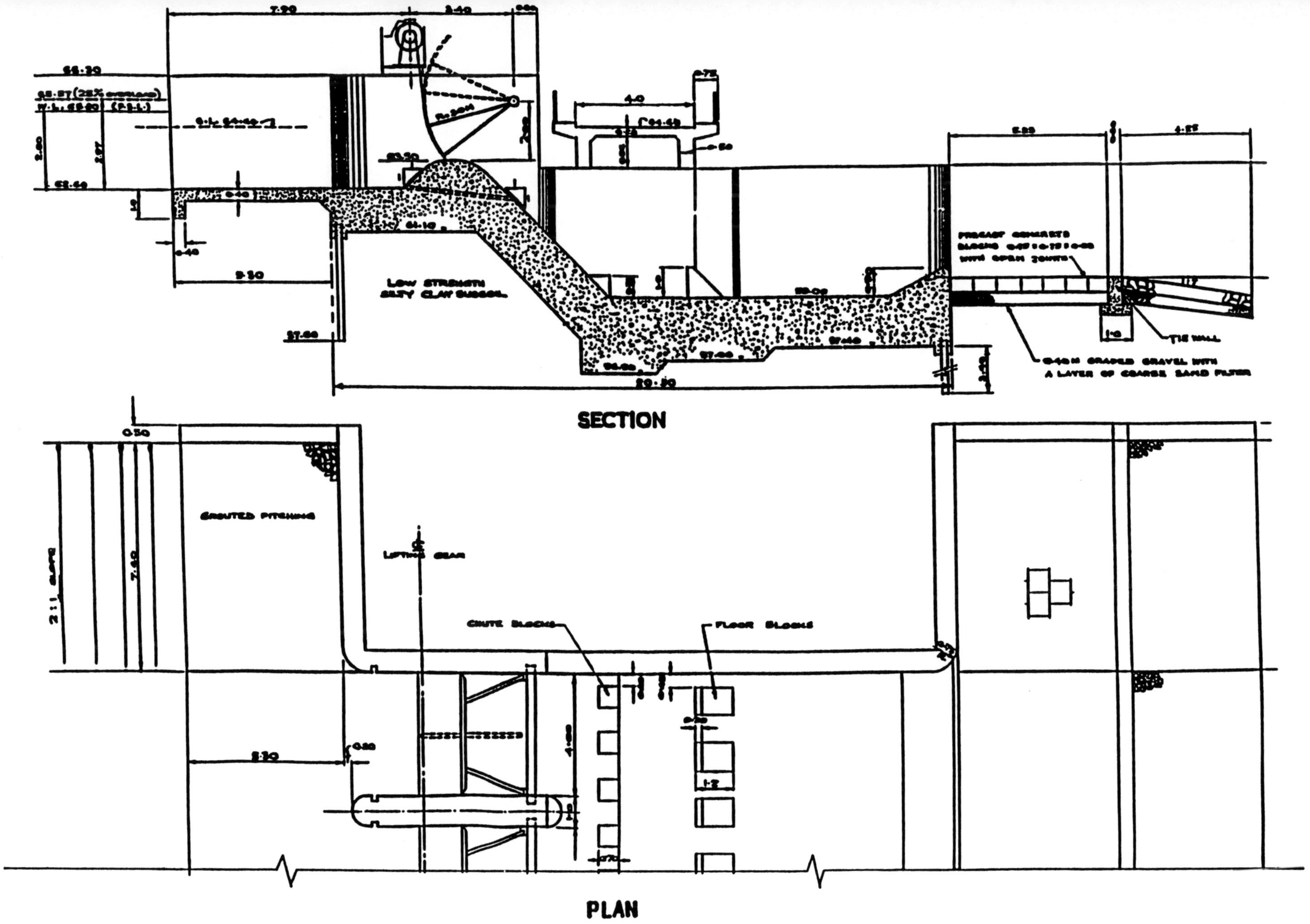

Fig.1 - Regulated Drop Structure

upstream protection in the form of stone pitching, an essentially curved weir with a short flat crest providing seating for gates in regulated falls, joined to a 45 degree inclined glacis, ending in a horizontal stilling basin with appurtenances to intensify energy dissipation in the hydraulic jump occuring under design conditions. This is followed by a talus protection consisting of concrete blocks overlying an inverted filter and stone pitching ending with a lip wall on the bed further downstream. The reverse filter acts as a safeguard against excessive uplift pressures downstream of the stilling basin.

The particular advantages of this type of design are that its performance is satisfactory over a wide range of discharges and drops, it does not require a long flat glacis and apron which makes the design more expensive and the cost of maintenance of an unflumed structure is likely to be less than a flumed one.

Materials and Construction

Mass concrete should be used in the design of drop structures as far as possible. Its use simplifies construction and is economical of upkeep and maintenance. All materials required for mass concrete construction and maintenance are available locally. Mass concrete hydraulic structures are durable and hard wearing, free from vibration and resistant to the abrasive and impact forces of water.

Reinforced concrete is used where the soil capacity is too low for the foundation pressure imposed by mass construction structures or where the use of mass concrete would be impractical or uneconomic compared with reinforced concrete, e.g., as in access road bridges spanning over falls.

Where regulation of canal drop structures is required steel crest gates will be included and of a design that can be fabricated locally. Similarly, other structure steelwork items such as footbridges, operating platforms and road bridge railings should be standardized of a type that can be made locally.

Gates for Regulated Structures

The types of gates normally used are radial gates. This type has been selected as it requires less maintenance than lifting gates, the operating gear is smaller and the superstructure is lower and less vulnerable to damage. The absence of gate grooves improves the hydraulic characteristics of the falls and effective sealing of the gates is not difficult.

The number of such gates required depend on the spans. It has been found that gates having standard spans of 4 and 3m are suitable for regulated main irrigation canal

falls. Factors affecting the selection of these standard gate spans are:

a)   Gates with spans restricted to 4 metres or less with span/depth ratios about 2:1 are of a convenient size for fabrication locally, transport handling and erection. They are also suitable for manual operation by one gatekeeper and do not require to be counterbalanced.

b)   The gated area is approximately 0.8 times the canal cross sectional area.

c)   Gates are designed to hold up an upstream water level equivalent to a 25% overload in the main canal and to be capable of passing a 50% overload on the design discharge.

Criteria for Weir Crest Levels

The following are the design criteria used in respect of surface flow and energy dissipation consideration:-

a)   The structure is to be unflumed, that is the weir is to span the full bed width of the canal in the case of a non-regulated (ungated) fall whilst in the case of a regulated fall it is to span approximately the full bed width to suit the number of standard gates incorporated in the structure.

b)   The required fall in water level is to be maintained through virtually the whole range of flow.

c)   The maximum height of the drop in water level is to be restricted to 3 meters.

d)   The structure is to be capable of passing an overload of 25% on the design discharge from the corresponding upstream water level, i.e., a peak discharge of 1.25 Q where Q is the canal design discharge.

e)   The structure is also to be capable of passing an overload of 50% on the discharge (1.5 Q) from a water level equivalent to the design discharge. This is to ensure that a structure will not require modification should a canal's capacity have to be increased at any time by widening.

f)   The implications of d) and e) above in the case of a non-regulated fall is that the selected weir crest level should be such that the structure is capable of passing either a discharge of Q or a flood discharge of 1.25 Q without any draw-down whichever condition gives the lowest crest level.

In this case an overload of 50% on the design discharge could only be passed by lowering the crest. This could be carried out without undue difficulty should a canal ever have to be remodelled.

In the case of a regulated fall the crest level selected is such at 1.5 Q can be passed from a water level corresponding to design discharge. In this case drawdown is prevented by operation of the crest gates which are able to hold up water to water level corresponding to peak discharge 1.25 Q.

Design for Uplift

As water flows through the soil under a regulator or a weir, gradually it loses head due to soil resistance until at the exit the head is reduced to zero. Thus, there will be an uplift pressure acting on the underside of the structure, and a certain exit gradient at the downstream end where water emerges from the soil. If at this end the upward force of the water is greater than the submerged weight of the soil particles they will be washed away: this is a dangerous condition which eventually may lead to the failure of the structure.

If follows that there are two essential conditions in the subsoil design of a hydraulic structure. For stability of the structure against uplift, a suitable floor thickness must be provided at different sections such that the weight balances the pressure force at each section. To safeguard against undermining the exit gradient must not exceed a certain safe limit generally 0.2 to 0.15. This is achieved by providing a suitable depth of pile line or cutoff wall at the downstream end of the floor.

For simple structures, theoretical solutions can be found for the distribution of uplift pressure and the value of the exit gradient. However, for most practical designs the boundary conditions are so complex that the theoretical equations are too difficult, and often impossible to solve. To overcome this difficulty, Khosla developed a method in which the theory for a flat floor with a pile at the downstream end is used in conjunction with empirical equations and charts. The method is known as the method of independent variables.

Besides this method, hydraulic and electrical models can be used to determine the flow lines and the distribution of uplift pressure. The electrical analogue may be conveniently used in the design office for analyzing structures resting on confined permeable foundations of complex geometry. These will be discussed in Chapter 23.

Channel Transition Structures

Introduction

The design of hydraulic projects often involves changes in shape and cross-sectional area of the stream whereby the uniformity of the flow is locally disturbed. A structure which is designed to reduce this disturbance and also to conserve the energy of the flow is called a transition, Fig.1. Where the flow is converging its purpose is to minimise head losses and where the flow is diverging its purpose is to recover as much as possible of the velocity head. Besides the hydraulic performance, transition design is equally governed by the economic importance of the structures.

Transitions are incorporated in the design of canal flumes, channel contractions and expansions, flumed falls, inlets and outlets for aqueducts, culverts, siphons and tunnels. Transition flumes are often combined in the design of other canal structures with economic advantage. For example, where a channel is to be crossed by a bridge, it is possible to constrict its width by a suitable transition (Fig.1) to keep the construction cost of the bridge to a minimum. Also, in the improvement of natural streams and channels to carry greater discharge, it is more economical to include a constriction transition under an existing bridge than to widen the bridge span.

Basic Characteristics

In transitions for subcritical flow, a minimum of head dissipation consistent with economy of construction is usually the guiding principle of design. The following assumptions are in order under these conditions:

a)   The gradient of total head is constant over the transition length, or the change in total head can be evaluated from Manning's equation for uniform flow by a step procedure.

b)   The velocity varies primarily as a function of distance.

c)   Curvature effects are negligible, the pressure distribution is hydrostatic, and separation does not occur.

A valuable tool in the analysis of such transitions is found in the specific-head curve; a series of these curves may be prepared for the given discharge and the various cross sections involved in a particular design. The curve can be plotted for a rectangular channel section in which the geometric change in channel shape is confined to changes in width, resulting in separate case, the area is the product of depth and width, whereas in general, the area is a more complex function of y depending upon the geometry of the local cross section.

The following information is usually available for the design of subcritical transitions:

a)   The geometric shape of upstream and downstream channels.

b)   The discharge with corresponding values of depth and area for upstream and downstream channels.

c)   The elevation of the total-head line for the upstream section and, with estimated adjustments, for all sections throughout the transition.

d)   The geometry of transition as dictated by the economic importance of the structure, and hence the specific-head curve for any immediate cross section.

T. H. Line

0·80m

1·40m

▽

Bed

LONGITUDINAL   SECTION

Inlet
Transition

Throat
Section

Outlet
Transition

3·8m

19m

9m

2·8m

PLAN

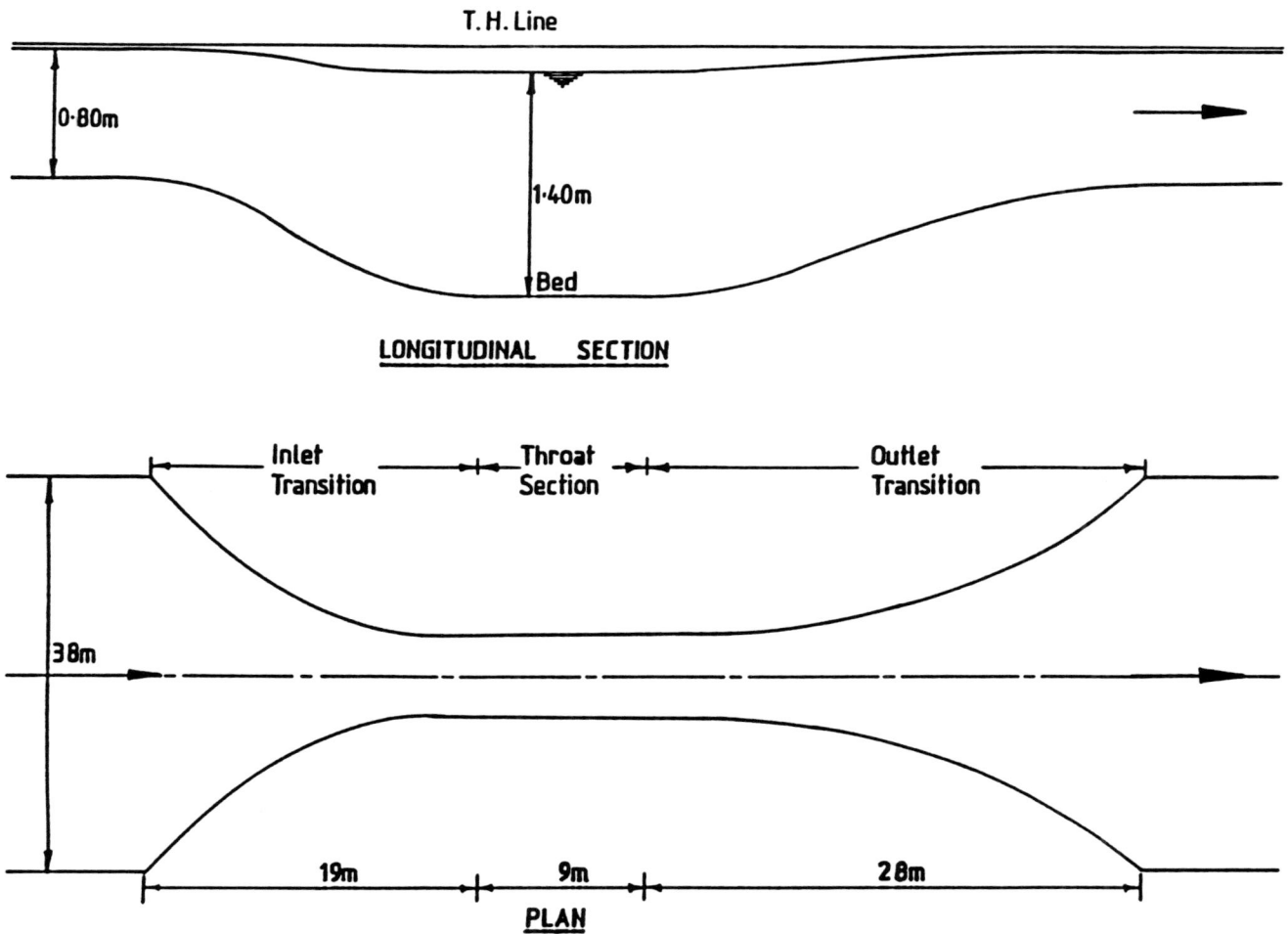

Fig 1 - Design of a bridge flume

Forms of Transitions

Under normal design and construction conditions, practically all canals and flumes require some type of transition structure to and from the waterways. The form of transition may vary from straight-line to very elaborate streamlined warped structures. There are four types of transitions (Fig.2) from trapezoidal to rectangular channels:

a)   Sharp transitions.
b)   The angular or cylinder-quadrant.
c)   The wedge type of transitions.
d)   The warped transition.

The first two are naturally not recommended for high velocities and the application of all four is restricted to subcritical flow.

Methods of Design

With the use of the specific-head diagram, three methods of transition design are possible:-

a)   Sketching Sidewalls - in this method the wingwalls of the structure are sketched in between the two reaches of the channel and then the bed profile is designed therefrom.

b)   Sketching Bed Profile - in this method the bed is sketched in, and then the wingwalls or sidewalls are designed therefrom.

c)   Sketching Water Surface - in this method the profile of the water surface is sketched in and then the wingwalls and bed are designed therefrom. In this method a trial-and-error

Cylindrical Transition

Wedge Transition

Warped Transition

Sharp Transition

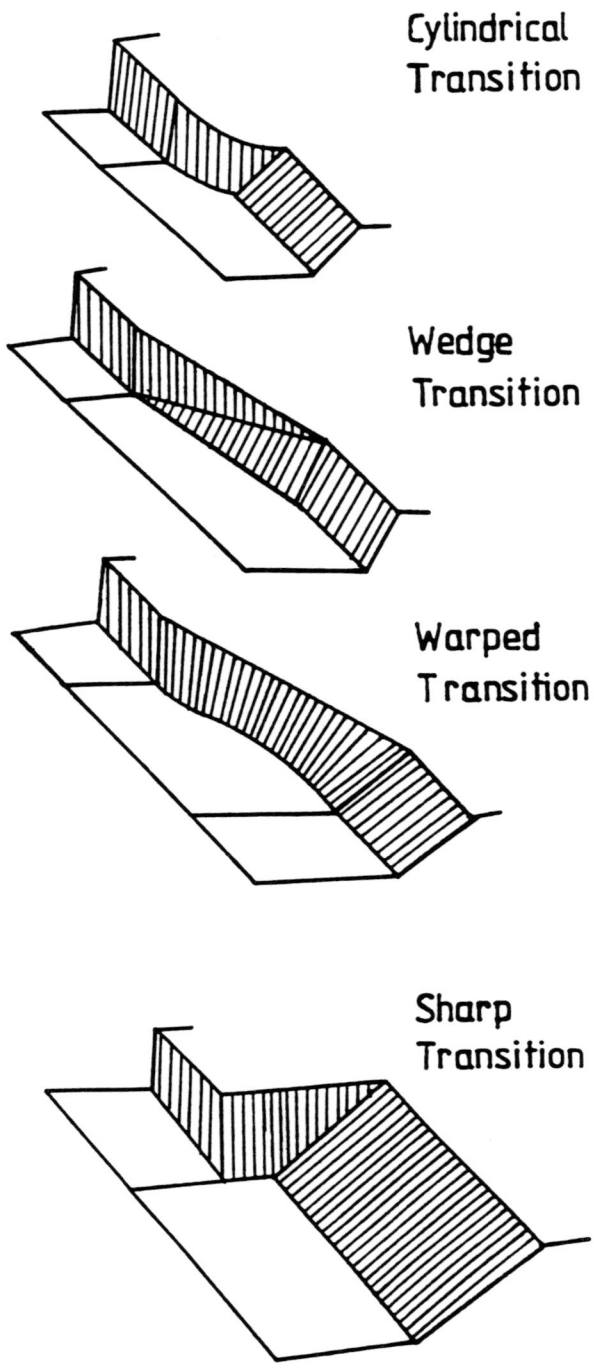

Fig 2 - Types of transitions

procedure is necessary for finding the values of y and H from the energy diagrams knowing only the velocity head.

d)   Sketching Sidewalls and Bed Profile - this method is used for syphon designs in which the wingwalls and bed are sketched in and the water surface and position of the roof is drawn.

The first method is most widely used and it usually gives the best results. To avoid the trial-and-error procedure of the last method, it is necessary to combine methods (b) and (c) a process which usually cuts the amount of work.

All methods require re-adjustments, therefore, until smooth lines are obtained by changing the longitudinal spacing for given geometric shapes of cross section, by adjusting the geometry itself, and finally by alternately adjusting bottom and surface elevations. The order in which the four forms are stated may also be taken as that of relative economy. Especially in structures of small size, the cylindrical or or circular transition walls are the most suitable froma construction point of view. In large structures the wedge-type transitions, and finally the warped transition, become almost as economical, since, quite aside from their hydraulic advantages, large surfaces facilitate construction. The design procedures for subcritical transition structures are described in Chapter 7.

16

DAM SPILLWAY STRUCTURES

Introduction

Usually the most important aspect of design of a dam is its spillway. The cost of a spillway is one of the major items of expense in a dam construction project and may well determine the type of dam selected. The function of the spillway is to provide an efficient and safe means of conveying flood discharges, exceeding the capacity of the reservoir to retain them past the structure to the downstream channel. Ordinary releases are handled through the regular outlet conduits so that the spillway may actually be used for high discharges. The capacity is determined by hydrological studies over the drainage area in question; in general it is desired to provide adequate capacity to transmit the largest flood that may possibly occur at the site. This will involve extrapolation of past flood frequency records and hydrological studies which will indicate the maximum rainfall and run off that can be produced on the area.

Although spillways are normally thought of in relation to dams, similar structures are used also for other purposes. In general, a spillway-type structure is required whenever it is necessary to convey a flow rapidly, safely and efficiently from a high to a low elevation, as in river weirs and canal falls and other drop structures. Such structures are frequently used in connection with irrigation, highway drainage and other hydraulic engineering projects.

Overflow Spillway Profiles

Engineers usually pattern the spillway crest after the nappe profile of the sharp-crested weir, in order to obtain the maximum delivery without reducing the pressure below that of the surrounding atmosphere (Figs 1 and 2). Theoretically, the adoption of such a profile, generally known as the Bazin profile, should cause no negative pressure on the crest. Under actual conditions, however, there exists friction due to roughness on the surface of the spillway. Hence, negative pressures on such a profile cannot be avoided. Consequently, the Bazin profile has been variously modified and many other profiles for design purposes have been proposed, among which are the Creager Profiles.

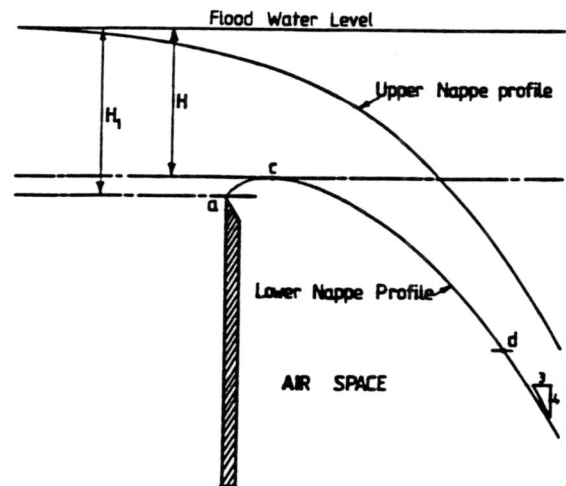

Fig 1 - Nappe profile for sharp crested weir

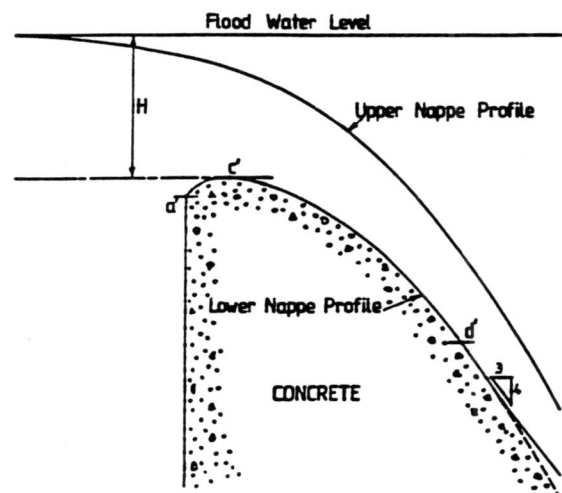

Fig 2 - Weir crest profile

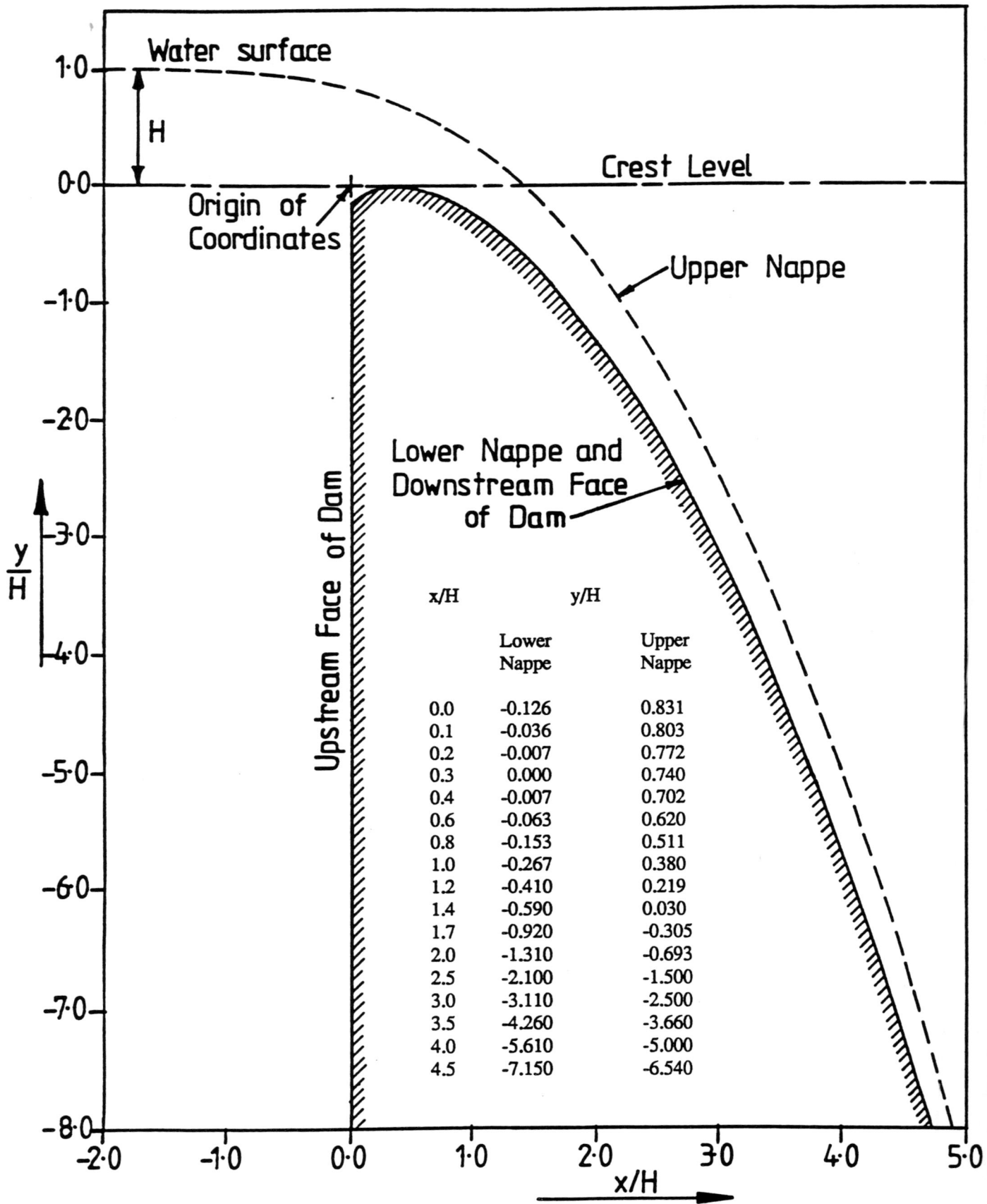

Water surface

1·0

H

0·0 ——— Crest Level

Origin of Coordinates

Upper Nappe

-1·0

Upstream Face of Dam

-2·0

Lower Nappe and Downstream Face of Dam

-3·0

x/H	y/H	
	Lower Nappe	Upper Nappe
0.0	-0.126	0.831
0.1	-0.036	0.803
0.2	-0.007	0.772
0.3	0.000	0.740
0.4	-0.007	0.702
0.6	-0.063	0.620
0.8	-0.153	0.511
1.0	-0.267	0.380
1.2	-0.410	0.219
1.4	-0.590	0.030
1.7	-0.920	-0.305
2.0	-1.310	-0.693
2.5	-2.100	-1.500
3.0	-3.110	-2.500
3.5	-4.260	-3.660
4.0	-5.610	-5.000
4.5	-7.150	-6.540

$\frac{y}{H}$

-4·0

-5·0

-6·0

-7·0

-8·0

-2·0   -1·0   0·0   1·0   2·0   3·0   4·0   5·0

x/H

Fig 3 - Dam spillway crest for vertical upstream face

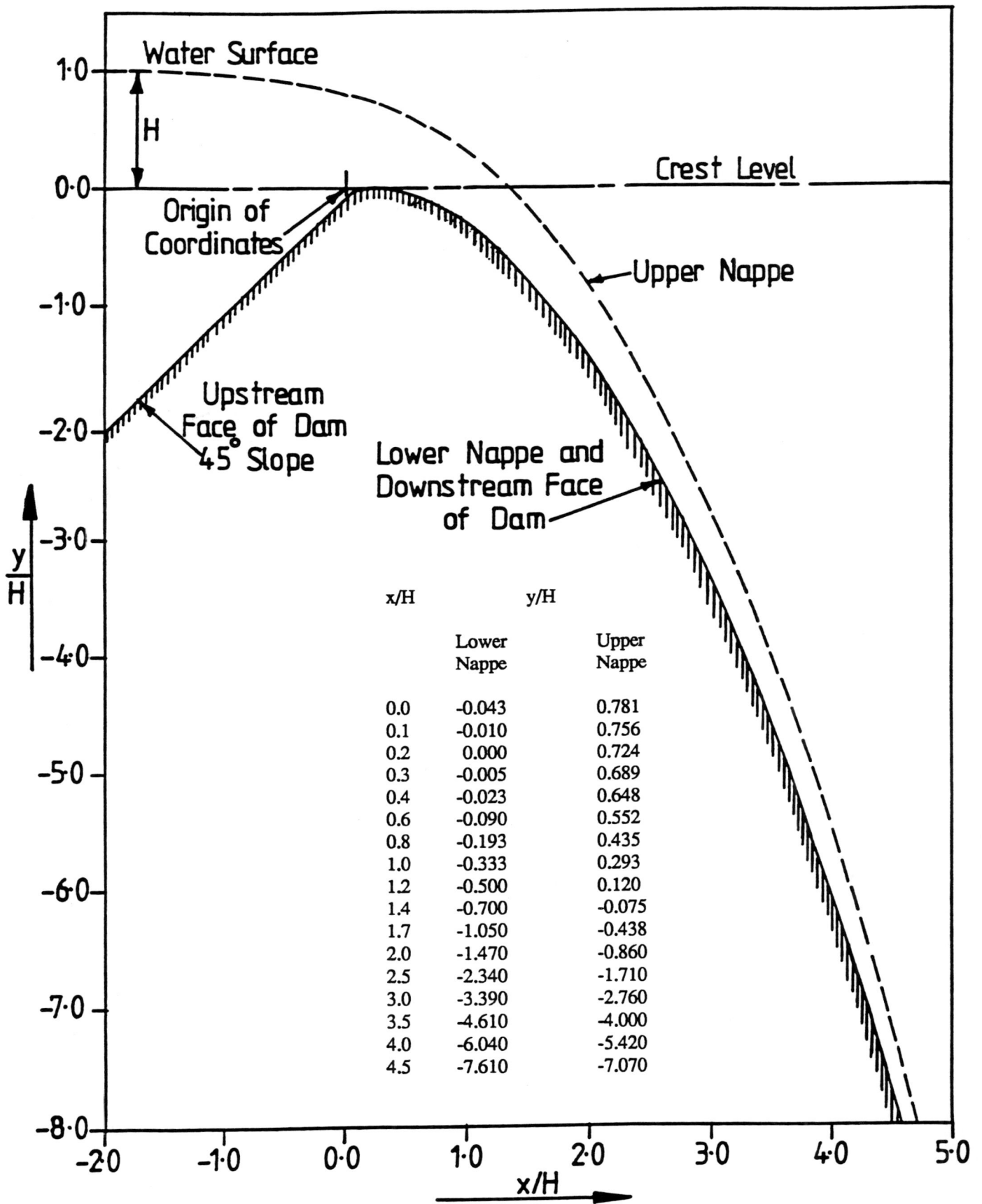

The figure shows a dam spillway crest diagram with the following labeled elements: Water Surface, Crest Level, Origin of Coordinates, Upper Nappe, Upstream Face of Dam 45° Slope, Lower Nappe and Downstream Face of Dam, and the dimension H between the water surface and crest level. Axes are labelled y/H (vertical) and x/H (horizontal).

x/H	y/H	
	Lower Nappe	Upper Nappe
0.0	-0.043	0.781
0.1	-0.010	0.756
0.2	0.000	0.724
0.3	-0.005	0.689
0.4	-0.023	0.648
0.6	-0.090	0.552
0.8	-0.193	0.435
1.0	-0.333	0.293
1.2	-0.500	0.120
1.4	-0.700	-0.075
1.7	-1.050	-0.438
2.0	-1.470	-0.860
2.5	-2.340	-1.710
3.0	-3.390	-2.760
3.5	-4.610	-4.000
4.0	-6.040	-5.420
4.5	-7.610	-7.070

Fig 4 - Dam spillway crest for inclined upstream face

Figures 3 and 4 give the recommended co-ordinates of the nappes and shape of the crest for dams with vertical face and with face inclined on a 45 degree slope, respectively and with no velocity of approach. The co-ordinates in the tables are to be multiplied by H to get the co-ordinates of the concrete crest. For appreciable velocity of approach, the shape of the crest can be approximated safely, for preliminary designs, by multiplying the co-ordinates in the table by Ho which includes the velocity head.

Other Weir Profiles

On the basis of the US Bureau of Reclamation data, other spillway shapes have been developed and represented by the following equations:-

$$X^n = K H_d^{n-1} Y$$

or

$$Y = \frac{X^n}{K H_d^{n-1}}$$

where X and Y are co-ordinates of the crest profile with the origin at the highest point of the crest, H is the design head excluding the velocity head of the approach flow, and K and n are parameters depending on the slope of the upstream face. The values of K and n are given below:-

Slope of Upstream face	k	n
vertical	2.00	1.850
3 on 1	1.936	1.836
3 on 2	1.939	1.810
3 on 3	1.873	1.776
(1 on 1)		

Discharge Over Spillways

Overflow spillways are normally used for measuring outflows from reservoirs. The discharge over a spillway can be computed by an equation in the form:-

$$Q = C B H_o^{1.5}$$

Where Ho is the total head on the crest, including the velocity head. Model tests of the spillway have shown that the effect of the approach velocity is negligible when the height h of the spillway is greater than 1.33 H where H is the design head excluding the approach velocity. For critical flow over the crest the coefficient of discharge can be shown to be C = 1.7.

Effect of Piers in Gated Spillways

Piers are needed to form the sides of the gates in gated spillways. The effect of the piers is to contract the flow, and, hence, to alter the effective crest length of the spillways.

The effective length of one bay of gated spillway may be expressed as:-

$$B = B_o - K N H_o$$

where Bo is the clear span of the gate bay between the piers; K is the pier contraction coefficient; N is the number of side contractions, equal to 2 for each gate bay; and Ho is the total head on the crest including the velocity head. The discharge coefficient however, is assumed to be the same in both gated and ungated spillways.

The pier contraction coefficient varies mainly with the shape and position of the pier nose, the head condition, the approach depth of flow, and the operation of the adjacent gates. The approximate K value ranges from 0.1 for thick blunt noses to 0.04 for thin or pointed noses and is 0.035 for round noses. These values apply to piers having a thickness equal to about one-third the head on the crest when all gates are open. When one gate is open and the adjacent gates are closed, these values become roughly 2.5 times larger.

Model Tests of Overflow Spillways

The profile of an overflow spillway can be designed for one head only. This head is the design head which generally produces a lower nappe of flow that agrees closely with spillway profile. The spillway, however, must also operate under other heads, either lower or higher than the design head. For lower heads, the pressure on the crest will be above atmospheric but still less than hydrostatic. For higher heads, on the other hand, the pressure will be lower than atmospheric, and it may become so low that separation in flow will occur. Model experiments indicate, however, that the design head may be safely exceeded by upto 50%, beyond this, harmful cavitation may develop. It is recommended that model studies are carried out for all spillways to check the preliminary design and to provide a rating curve for measurement.

PART III

CONFINED JETS
AND STREAMS

Surface Plane Jet
Parallel Wall Jet
Deflected Round Jet

SURFACE PLANE JET

Introduction

The jet from an overflow weir or from the orifice between the gates of a regulating sluice on entering the pool of water downstream, either flows along the surface, or alternatively is deflected downwards becoming fully submerged and sweeps along the bed of the pool, Figure 1. Both flows seem fairly stable; intermediate ones seldom occur. When the flow is reduced by increasing the downstream water level, submerged flow may suddenly change over to surface or wavy motion and vice versa.

A schematic section of the surface motion of the jet is shown in Figure 2. The jet flows over the slow moving fluid in the pool and is confined by a lower boundary. The flow consists of two parts: the potential core near the inlet enclosing a prism of fluid in which the velocity is uniform and equal to the initial value U_o, and the fully developed jet where the surface stream and the slow

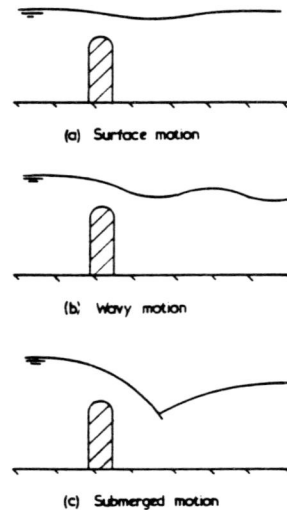

Figure 1 – Types of Flow Downstream of a Weir

(a) Surface motion

(b) Wavy motion

(c) Submerged motion

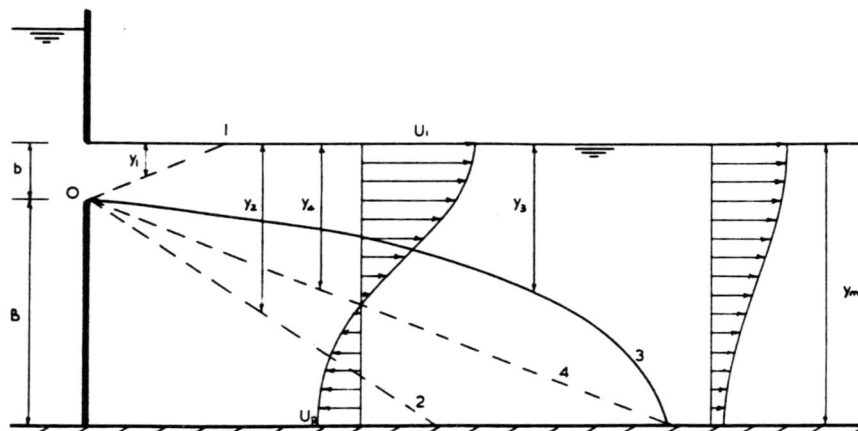

Figure 2 – Surface Motion of a Plane Liquid Jet

moving fluid in the pool below form a turbulent mixing region with an inner boundary 01 and an outer boundary 02. In the latter region, the velocity profile consists of a mixing layer with a maximum velocity U_1 at the water surface and a counterflowing layer with a maximum reverse velocity U_2 near the bed of the pool.

The mixing process between the surface stream and the surrounding fluid causes part of the latter to be carried forward with the jet under conditions in which both the forward momentum and the total discharge are conserved. The process of entrainment leads to recirculation to replace the fluid entrained and a zone of circulation motion is established underneath the surface stream. The line 03 defines the meanflow line bounding the fluid which is sucked outward from the stream.

The behaviour of such a jet is of importance in the design of partially submerged orifices, regulators, weirs, sedimentation tanks and fish passes of the pool type; yet very little information has been published on the flow. In their investigation of fish passes, Nemenyi and White [1] observed the flow over a weir and found that the surface profile had very little effect upon the change from surface to submerged motion, and that the principal factor was the thickness of the weir. The present research was undertaken partly to provide information for the design of stilling basins downstream of moveable canal regulators, and partly to lead a better understanding of the complex behaviour of confined jets and streams under investigation by the author [2 to 8].

Analytical approach

The physical aspects of dispersion of a surface jet has similarity with the mechanism of a plane submerged jet and the confined flow in the circulation zone behind the two-dimensional bluff body. A knowledge of the behaviour of such flows is therefore of importance in explaining the more complex motion of the surface yet. For the plane jet, Tollmien [9] found that in the region of potential core the inclination of the streamlines forming the limits of the jet are: (i) boundary on the uniform stream side expands at an angle of 1 in 12; (ii) boundary on the fluid at rest side expands at an angle of 9.8°. In the region of fully developed flow, half of the width of the mixing zone is given by:

$$b = 0.266 \ x$$

or with an angle of expansion of about 15° or 1 in 4. The central velocity at a distance x from the plane of the opening is given by

$$\frac{U}{U_o} = 3.64 \ \sqrt{\frac{b}{x}}$$

where U_o is the jet velocity at exit.

Abramovich [10] analysed the flow behind a two-dimensional bluff body and found the length of the circulation zone to be 6.1B where B is the half width of the body and in the case of the surface jet it is the depth of the pool. This theory is correct as long as in the stream the core of constant velocity is retained throughout the length of the circulation zone, and as this core decays in a shorter distance, the theory becomes less accurate.

Experimental conditions and methods

The experiments were conducted in a standard laboratory glass flume, 300mm wide by 300mm deep and 10m long, having an outlet weir for adjusting the depth of flow, and a weighing arrangement for measuring the discharge. The flume was fitted with two perspex sheets, each 25mm thick, to produce through a slit a surface water jet of depth b = 19 mm and width 300mm, as shown diagrammatically in Figure 2. The uniform velocity of the jet at exit was measured to be $U_o = 0.605$ m/s. The Froude number of the jet was calculated as

$$F = U_o/\sqrt{gb} = 0.605/\sqrt{9.81 \ x \ 0.019} = 1.4$$

The depth of the pool beneath the jet was B = 8b = 152mm and it extended 8m downstream of the jet exit. In comparison, the length of the recirculation flow in the pool was found to be about 6B = 912mm. All the results presented in this paper were deduced for this one jet, having the same density as the surrounding fluid.

The velocity distributions along the centreline of the flume were measured by a shielded dynamic probe [2]. The probe was found to be insensitive to direction changes through a range of angles of $\pm$ 40 degrees. The pressure distribution at any section of the flow was assumed hydrostatic and was locally measured by an accurate point gauge.

The velocities were also measured by a photographic technique developed by the author [11, 12]. For ease of photography, the experiments were carried out in a smaller glass flume, 100mm wide by 300mm deep and 3m long, other dimensions of the flow remaining the same as above. The flume extended 2.75m downstream of the jet exit. A mercury-vapour discharge lamp was used to produce a sheet of light 10mm thick by 200mm long along the centreline of the flume. By using a stroboscopic light, the paths of the illuminated oil tracers over successive periods of one-hundredth of a second were obtained as a series of dashed streaks, and by comparing the length between the centres of adjacent dashes with a linear scale included in the photograph, the magnitudes of velocities were obtained.

Typical photographs of the various regions of the jet are shown in Plates I, II, and III. The local depth of flow is 171mm and the flow is from right to left. The values of x/b approximately define the beginning and end of each photograph. A white scale is placed near the downstream end just above the water surface. The results obtained using such photographs were practically the same as those measured by the shielded probe. The reduction in width of flow from 300mm to 100mm apparently did not influence the distribution of velocities in the jet.

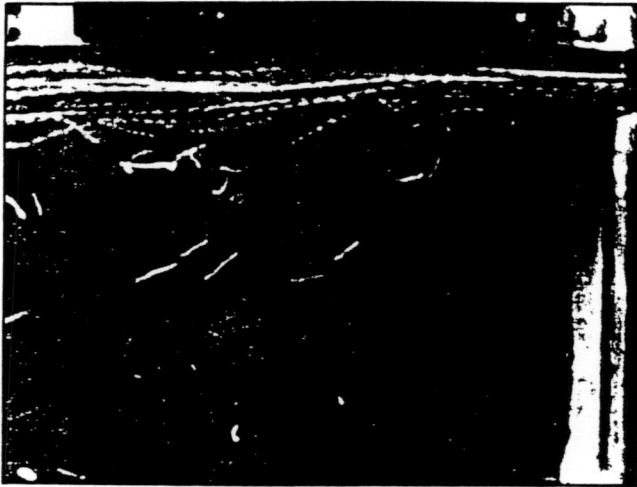

Plate I – Flow Near Jet Exit, x/b = 0-12, Exposure 0.25 Second.

Plate II – Flow in Entrainment Region of Jet, x/b = 12-24, Exposure 0.25 Second.

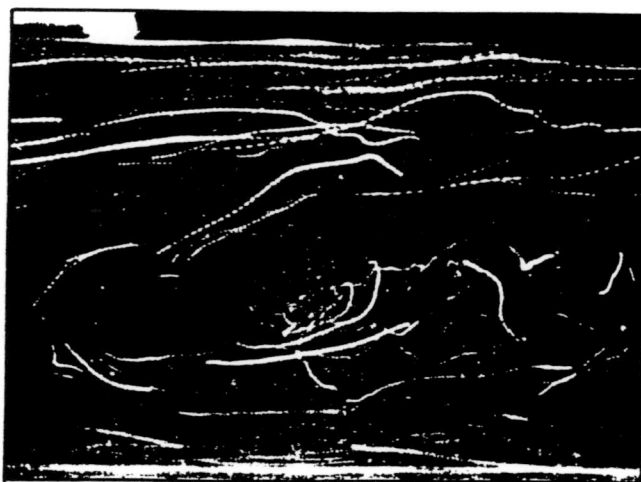

Plate III – Flow in Recirculation Region of Jet, x/b = 30-42, Exposure 0.5 Second.

Experimental results

Regions of Flow

The results show that four regions are recognised along the direction of flow of the jet. These regions are mainly defined in the rate of decay of maximum stream velocity that exists just below the water surface, Figure 3. Region I of potential core extending to 8 widths from the outlet face, where the maximum velocity is constant and is equal to the original outlet velocity U_o. This distance is 4 widths shorter than that for the two dimensional free jet.

Region II of entrainment flow that extends to 28 widths, in which the maximum velocity varies inversely with distance from the outlet as given by:

$$\frac{U}{U_o} = 4.09\sqrt{\frac{b}{x}}$$

This is practically the same as that for the free plane jet except that the rate of decay is slightly faster. Region III of recirculated flow that extends to about 48 widths in which the flow is sucked from the jet into the recirculation zone. The rate of decay of maximum velocity is variable due to the gradual rise in pressure in this region. Region IV of transformed flow which extends a long distance downstream and in which the expanded jet is transformed into a uniform rectangular stream for normal flow in an open channel. The rate of decrease of maximum velocity seems to vary inversely with $x^{1/2}$.

Jet Boundaries

Figures 4 and 5 show that the water surface profile and boundaries of the jet plotted non-dimensionally. Apart from a slight reduction near the outlet, the depth remains constant up to about 28 widths, the point where the outer boundary 02 touches the bed of the pool. Further downstream the level gradually increases until it reaches the depth of flow in the channel. The slope of the water surface indicates the conversion of the kinetic head of the jet into potential head in Region III of recirculated flow.

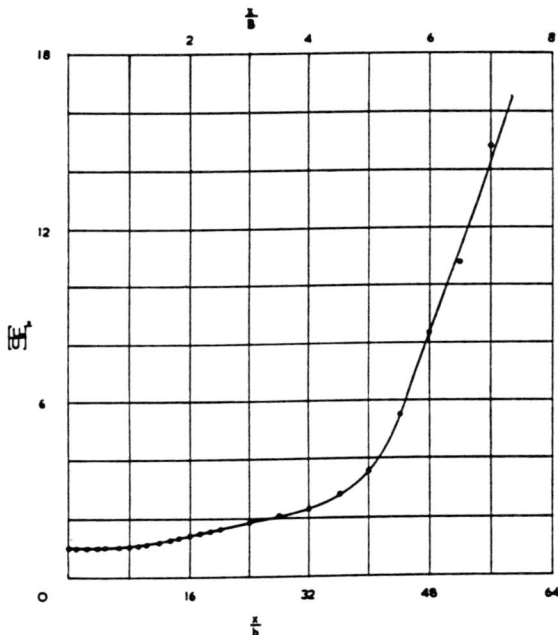

Figure 3 – Decay of Maximum Velocity

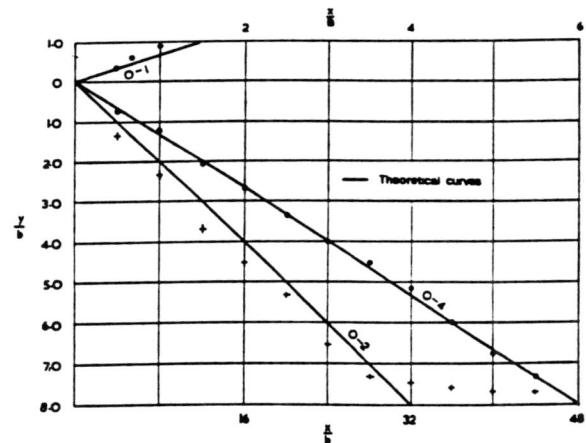

Figure 5 – Spreading of Jet Boundaries

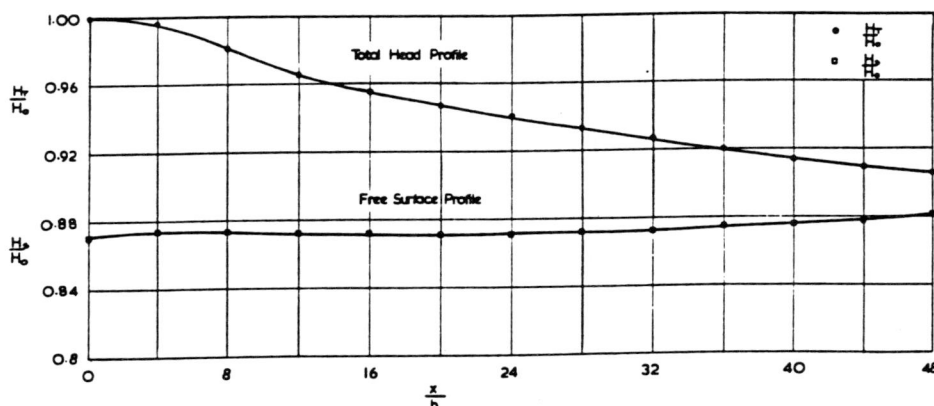

Figure 4 – Water Surface and Total Head Profiles

25

The inner boundary of the potential core is seen to expand at an angle of 6.4° or about 1 in 9, which is greater than that predicted for the boundary 01. The spreading angle of the outer boundary 02 is 15.3° and this is in agreement with Tollmien's value for the plane jet. The boundary 04 defined by u = 0, expands at an angle of 9.4°, again consistent with Tollmien's prediction. Where the boundary 03 reaches the bed a stagnation point is formed, marking the end of the circulation zone. The length of this zone is about six times the depth of the pool, compared with Abramovich's theoretical value of 6.1 B.

Velocity Distributions

The velocity profiles in various regions of the jet are shown in Figures 6 and 7. In the region of entrainment the profiles are similar and are in satisfactory agreement with Tollmien's curve. Further downstream the profiles are still largely similar; the actual velocity falls faster on the stream side but slower on the circulation side than given by the theoretical curve. This is undoubtedly due to the pressure rise along the surface of the jet and to the curtailment of jet boundary 02 in this region. The variation of the maximum reverse velocity in the backflowing stream is given in Figure 8. The velocity reaches a peak value of 0.4 U_o at $x/L \simeq 0.6$, the end of the entrainment region of the circulation zone. Due to the relatively short potential core, the results are not in agreement with Abramovich's curve. However, they can be approximated by a sine curve as follows:

$$\frac{U_2}{U_1} = 0.30 \sin\left(\frac{\pi x}{L}\right)$$

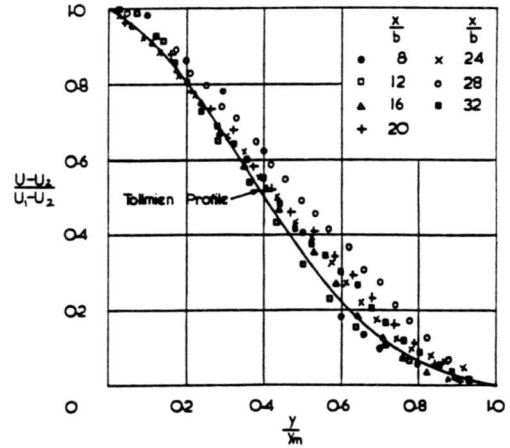

Figure 6 – Velocity Profiles in Region of Entrainment

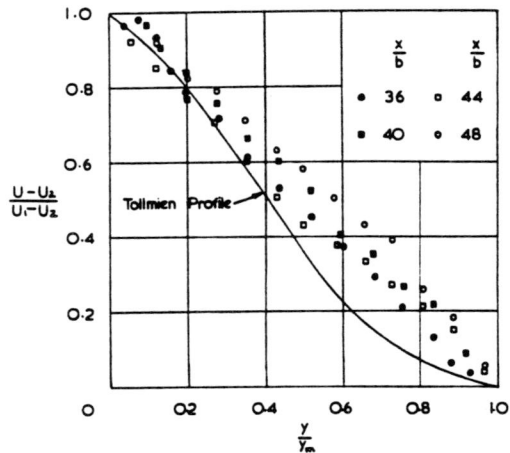

Figure 7 – Velocity Profiles in Region of Recirculation

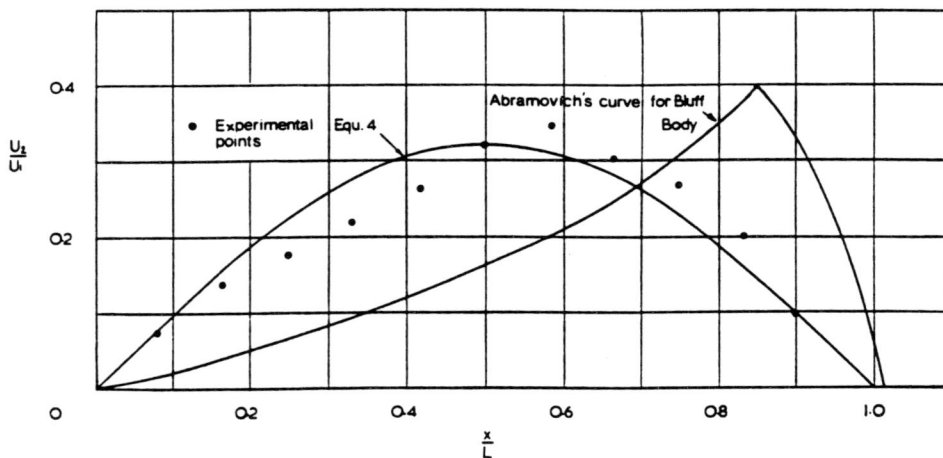

Figure 8 – Variation of Maximum Reverse Velocity

26

Further analysis led to the surprising results shown in Figure 9, where it is seen that experimental data closely lie to a Gaussian distribution curve represented by the following equation:

$$\left(\frac{U_2}{U_1}\right)^2 = \frac{K_1}{\sqrt{2\pi}}\, e^{-\frac{t^2}{2}} - K_2$$

where $K_1 = 0.227$, $K_2 = 0.008$ and

$$t = \sqrt{5\left(\frac{1-2x}{L}\right)}$$

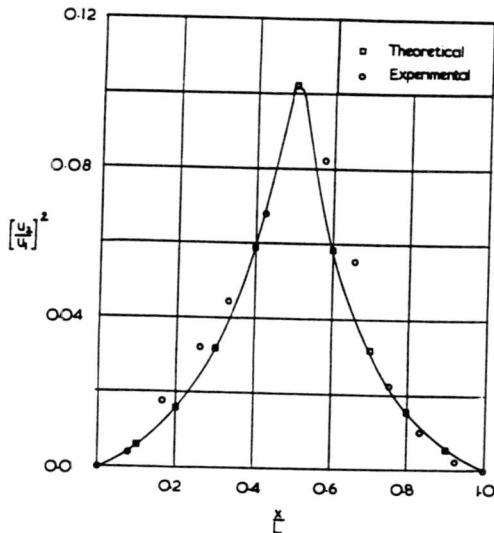

Figure 9 – Gaussian Distribution of Reverse Velocity

Total Head Characteristics

Due to the pressure rise in the region of recirculated flow it may be more appropriate to study the decay of the jet by considering the variation of total head along the direction of flow. At any vertical section of the mean flow, Figure 1, the total head is the sum of the static and kinetic heads and may be represented as follows:

$$H_T = H_s + H_k$$

or

$$H_T = y + \frac{1}{2gq}\int_o^y u^3\, dy$$

where y is the local depth of water and q is the unit discharge. Figure 10 shows the experimental results plotted as log (H_T/H_o + 1) against log (x/b) where H_o is the total head of the jet at inlet. Downstream of the potential core, the straight line graph clearly indicates that for both the entrained and recirculated regions of the jet, the variation of total head is governed by the same power law, despite the pressure rise in the latter region, as given by the following expression:

$$\left(\frac{H_T}{H_o} + 1\right) = 2.10\left(\frac{b}{x}\right)^{0.024}$$

It could therefore be said that for liquid jets diffusing under gravity the total head seems to be a more general parameter than the maximum velocity for studying the transformation of the jet within the whole of the circulation zone.

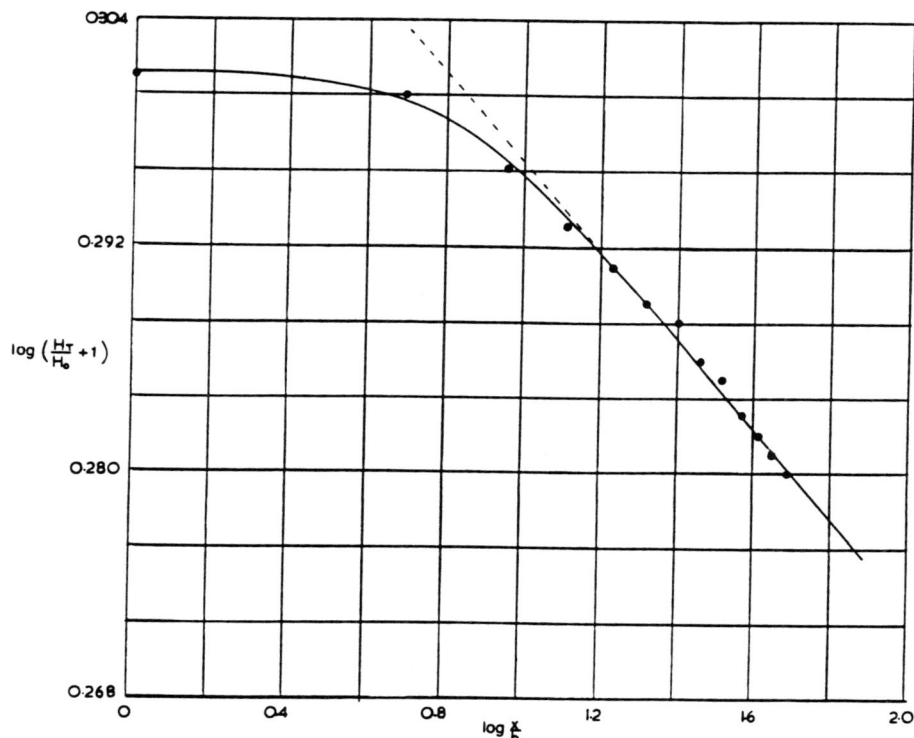

Figure 10 – Decay of Total Head

Another insight into the flow may be obtained by plotting the rates of static pressure head, dh_s/dx, of kinetic head, dhk/dx, and of total head, dH_T/dx, with distance as shown in Figure 11. For the static head, the rate is small and constant in the entrainment region, then gradually decreases until the stagnation point is reached. The curves for the kinetic and total head rates are similar; both rates increase steeply to a maximum value just past the end of the potential core. Further downstream the rates decreases shapely, the two curves interesting near the end of the entrainment region where the outer boundary of the jet (02) reaches the bed of the pool. In the region of recirculation, the rate of dissipation of kinetic head remains surprisingly constant, while the rate for total head decreases gradually following a similar trend to that for the static head in the same region. This clearly indicates that the rate of dissipation is a maximum in the first half of the zone of diffusion.

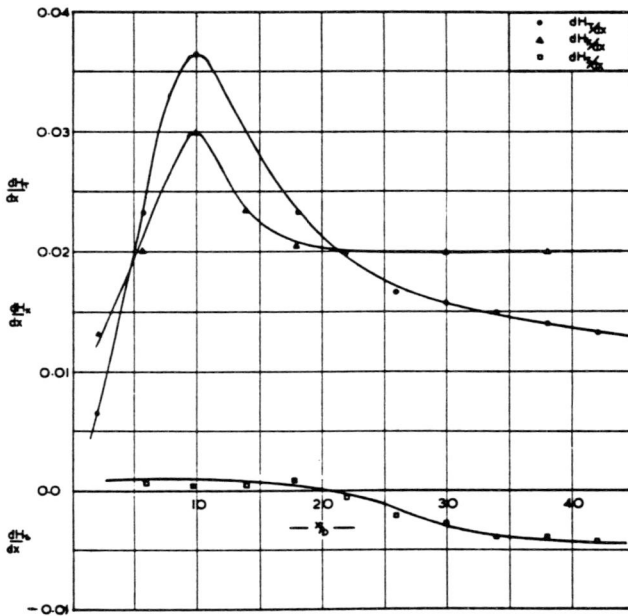

Figure 11 – Rates of Dissipation of Heads

References

1. Nemenyi, P & White, C.M. "Report on hydraulic research on fish passes : Appendix of Report of the Committee on Fish Passes". Inst. Civil Eng., London. 1948.

2. Naib, S.K.A. "Mixing of a subcritical stream in a rectangular channel expansion. J. Inst. Wat. E., Vol.20, No.3, p. 199, May 1966.

3. Naib, S.K.A. "Flow patterns in a submerged liquid jet diffusing under gravity. Nature, Vol.210, P.694, May 14, 1966.

4. Naib, S.K.A. "Unsteadiness of the circulation pattern in a confined liquid jet. Nature, Vol.212, p.753, November 12 1966.

5. Naib, S.K.A. "Abnormal grouping of large eddies in a submerged jet. La Houille Blanche, No.3, p.282, 1967.

6. Naib, S.K.A. "Stream boundaries of subcritical flow in a trapezoidal channel expansion. Wat. & Wat. Eng., Vol.73, p.155, April 1969.

7. Naib, S.K.A. "Spreading and development of the parallel wall jet. Aircraft Engineering, Lond. p.30, December 1969.

8. Naib, S.K.A. "Deflexion of a submerged round jet to increase lateral spreading. "La Houille Blanche, No.4. p.55, 1974.

9. Tollmien, W. "Calculation of turbulent expansion processes. N.A.C.A., TM No.1085, 1945.

10. Abramovich, G.K. "The theory of turbulent jets, M.I.T. Press, Boston, USA, 1963.

11. Naib, S.K.A. "Photographic method for measuring velocity in a liquid jet. The Engineer, Lond., Vol.221, p.961, June 24, 1966.

12. Naib, S.K.A. "Measuring velocity in a liquid jet. The Engineering, Lond., Vol.222, p.236, Aug.12, 1966.

PARALLEL WALL JET

Summary

This paper concerns an experimental investigation of a round air jet projected parallel to a wall. Experiments were carried out to establish the shape of the velocity profiles, the decay of maximum velocity and the rate of growth of the jet. The results are compared with Tollmien's theory for the free jet. All data are presented in a form readily available for design purposes.

Introduction

When a round air stream is projected parallel to a wall into the stationary atmosphere, the turbulent mixing causes the jet to entrain the surrounding air and spread out over the surface as shown in Fig.1. At the orifice, the stream has a constant velocity U_1. Between this stream and the adjacent atmosphere, a turbulent mixing region is formed having normal and lateral boundaries b and l. The normal velocity profile at any section consists of a wall boundary layer, a uniform stream extending to the centreline, and an outer jet layer. The mixing of fluid in the inner half of the jet may be expected to show some resemblance to the flow in a pipe or along a flat plate, whereas in the outer half of the jet the mixing of fluid with the surrounding atmosphere is likely to be similar in character to that of a round free jet discharging into stationary fluid.

Little work on the parallel wall jet has yet been published. Chester et al (Ref.1) carried out tests on air jets from a 2.5 cm. diameter orifice and from a 0.25 in. rectangular slot of the same area blown over a wooden surface set with its leading edge 1/16 in. below the edge of the aperture. Using a total-head tube, they measured the velocity contours in the jet and found that the angle of boundary divergence across the surface is 50 deg. while the normal boundary angle measured from the surface at centreline of jet to be 5 deg. Kuchemann (Ref.2) studied the behaviour of an exhaust jet in close proximity to a flat surface in

order to find out whether the axially symmetric stream tends to approach the surface. He measured the velocity profiles for large and short wall distances from the exit nozzle.

As a further contribution, the present investigation was undertaken to establish the shape of the velocity profiles, the decay of maximum velocity, and the spreading of the boundaries of the jet. Besides its application in certain fields of engineering design, it was hoped that the research will lead to better understanding of the behaviour of confined liquid jets and streams with circulation zones currently under investigation by the author (Refs. 3, 4, 5, 6).

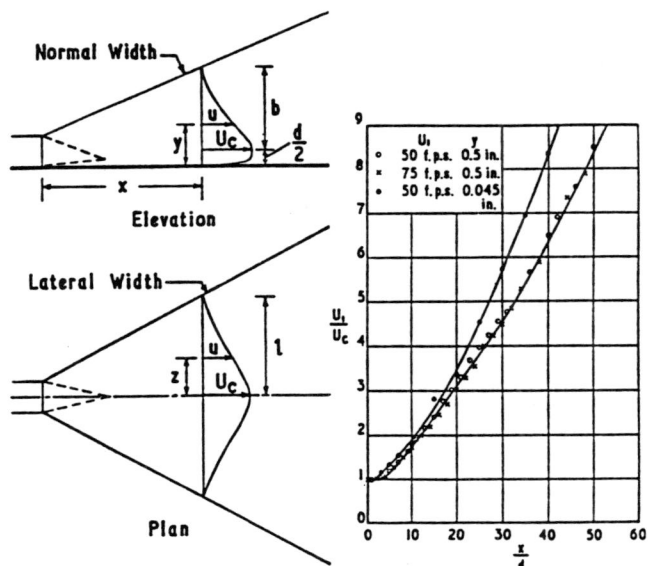

Fig. 1.—Flow in a parallel wall jet. Fig. 2.—Variations of maximum velocities.

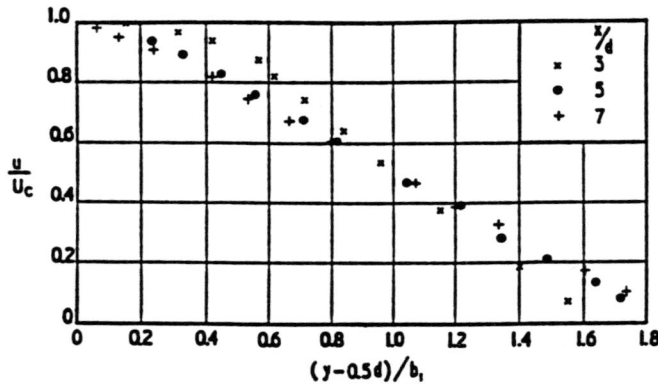

Fig. 3.—Velocity profiles in the transition region.

Analysis of Turbulent Round Jets

The spreading of a round jet in an infinite fluid is well investigated, and the results while not directly applicable are of great help in explaining the action of the present jet. For an excellent discussion of the subject, the reader is referred to Abramovich's book (Ref.7).

Using Prandtl's equation for turbulent shear stress, Tollmien (Ref.8) was the first to solve the problem of a free round jet originating from a point source. The theory establishes the shape of the velocity profile at any section and shows that the central velocity of the jet U_c is inversely proportional to the distance,

$$\frac{U_c}{U_1} = K \frac{d}{x}$$

The value of K has been found experimentally to be variable depending on the effect of the boundary layer in the nozzle. Based on theoretical and experimental results, Squire (Ref.9) recommends a mean value of 6.5.

The analysis also gives the outer boundary of the jet by the following:

$$b = 0.214 \ x$$

which makes the angle of divergence approximately 12 deg. with the jet axis.

Kuethe (Ref.10) extended Tollmien's work to the case of a jet with a finite source and established the length of the potential core is about 5 diameters from the orifice. This result agrees with experiments over a wide range of Reynolds numbers.

An interesting phenomenon arises when the free jet strikes a surface at right angles and spreads out radially over it. This flow has been termed a radial wall jet. Glauert (Ref.11) analyzed the flow and obtained 'approximate similar' solution in regions far downstream of the point of impingement. The analysis near the wall was carried out with the eddy viscosity based on Blasius law for pipe flow $y\alpha \ U^6$, while Prandtl's mixing length hypothesis was used further out.

An experimental investigation of the radial wall jet has been carried out by Bakke (Ref.12). Measurements of velocity profiles were made at distances 10 to 20 times the jet diameter. At these distances, the profiles were similar, and the rate of spreading and decay of the jet followed power laws:

$$b\alpha \ z^{1.02}$$
$$U_c \ \alpha \ x^{1.14}$$

which are in fair agreement with Glauert's predictions.

Apparatus and Experimental Method

A jet of air, 1 in. diameter, was produced by a blower with a rheostat for adjusting the speed of the fan. The jet was directed parallel to a smooth wooden surface 6ft long by 8ft wide with its leading edge set flush with the nozzle. Except for the small boundary layer on the inside of the nozzle, the velocity profile was practically uniform across the exit-section.

A shielded pressure probe (Ref.13), 0.25 in. external diameter, with central sting (o.d. of 0.075 in. and i.d. of 0.04 in.) was used for velocity measurements. The probe has been found to be insensitive to direction changes through a range of angles plus or minus 45 deg. for a 1

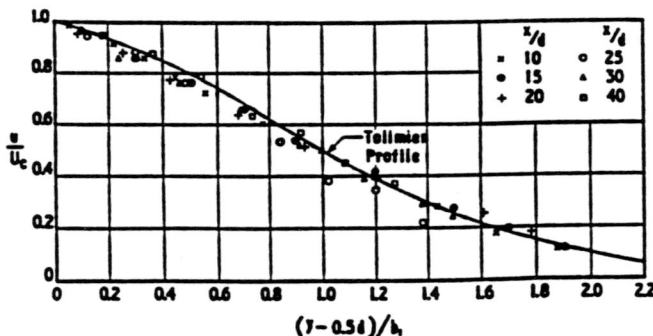

Fig. 4.—Normal velocity profiles in the outer half of the jet.

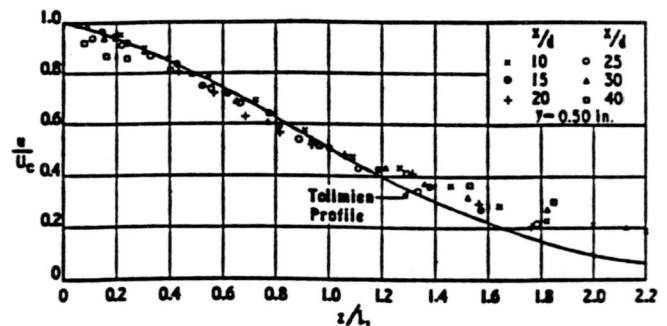

Fig. 5.—Lateral velocity profiles along the centre plane.

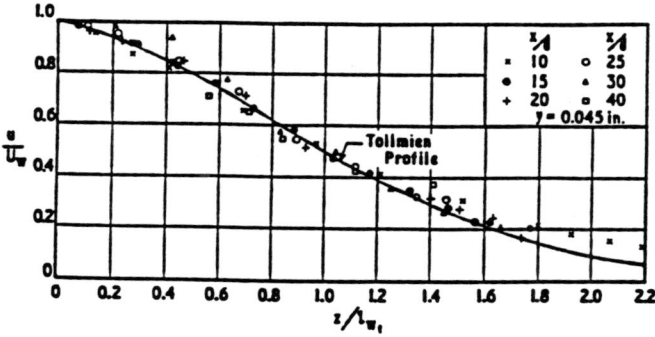

Fig. 6.—Lateral velocity profiles near the wall, y = 0.045 in.

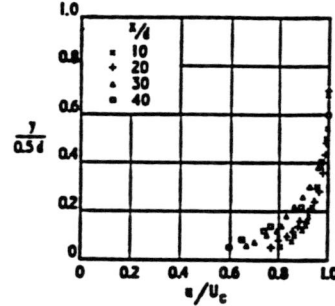

Fig. 7.—Normal velocity profiles in the inner half of the jet.

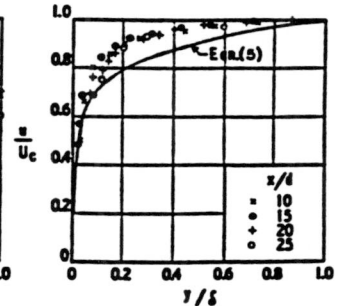

Fig. 8.—Velocity distribution in the boundary layer.

per cent error limit. Near the wall, y = 0.045 in., the velocity profiles across the surface were measured by a total-head tube 1mm. inside diameter. Each probe was clamped into a micrometer screw traversing mechanism graduated in 0.002 inch.

Decay of Maximum Velocity

The variations of maximum velocity near the wall and along the centreline of the jet are plotted non-dimensionally in Fig.2. Near the wall, the variation of velocity follows a curve. Along the centreline the velocity is seen to be constant over a certain distance followed by a transitional length which joins to a straight line and this is followed by another line of greater slope. The length of the potential core as given by the horizontal part of the graph is seen to be 2 diameters compared with about 5 diameters for the free jet. As expected, the presence of the wall on one side of the jet increases lateral mixing and hence reduces the length of the potential core. However, the length of the transition region is seen to be such that the linear distribution begins at a distance of 8 diameters as for a free jet.

The subsequent linear relationship between the reciprocal of the velocity and the distance is seen to hold for a distance up to 33 diameters and the results may be represented by the expression

$$\frac{U_c}{U_1} = \frac{7.0 \, d}{(x+2.5)}$$

which differs from the relation for the radial wall jet but it is practically the same as for the free jet, Eq. (1). The origin was located by extrapolating the straight line thus obtained to meet the x-axis. The data for the last region of flow lies on a straight line represented by

$$\frac{U_c}{U_1} = \frac{5.0 \, d}{(x-8)}$$

Thus, for distinct regions of flow are apparent in the outer half of the jet: (i) region A of potential core extending about two diameters from the orifice, (ii) region B of transition flow that extends up to 8 diameters and in which the velocity apparently varies parabolically with distance, (iii) region C of established flow which extends up to about 30 diameters and (iv) region D of terminal flow in which the residual velocity decays rapidly as a result of large scale turbulence to a value below 50 ft./min. commonly regarded as still air.

Non-dimensional Velocity Profiles

In order to compare the velocity profiles in various regions

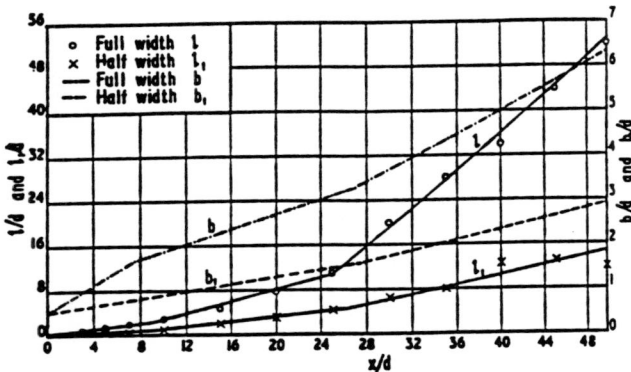

Fig. 9.—Spreading of the outer half of the jet.

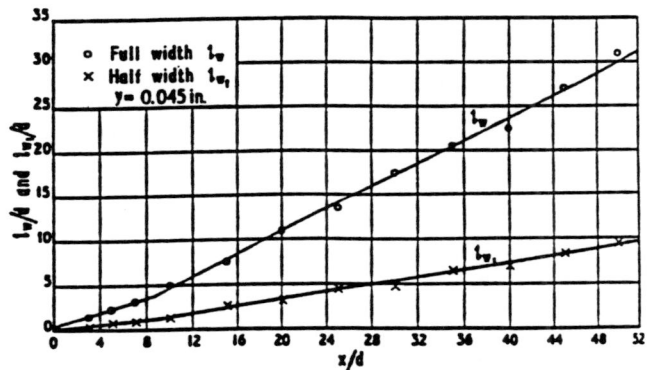

Fig. 10.—Spreading of the jet near the wall, y = 0.045 in.

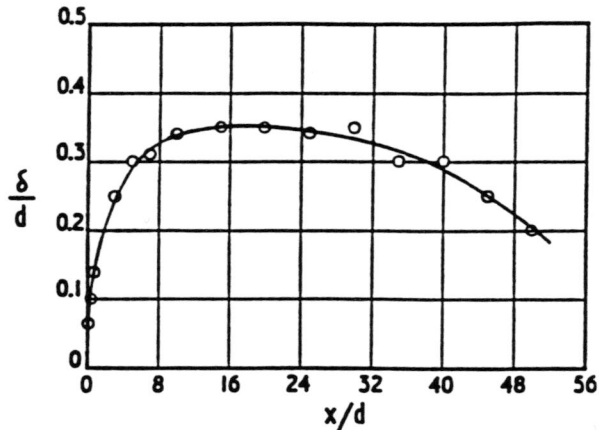

Fig. II.—Variation of boundary layer thickness.

$$\frac{u}{U_c} - \frac{(y)^{\frac{1}{7}}}{(\delta)}$$

It is seen that the points deviate appreciably from the curve, particularly near the middle of the layer. This is most probably due to the flattening effect of the wall on the jet.

Spreading of Boundaries

Fig.9 shows the widths b and l for the outer half of the jet drawn non-dimensionally against x/d. The boundaries are seen to increase linearly and show the different regions of flow indicated by the decay of maximum velocity. A noticeable effect is the rapid spreading of width b in the vicinity of the orifice.

The angles of normal and lateral boundary divergence θ and ϕ in the regions of fully developed flow are given below.

Boundary	Angles of Divergence	
	Region C	Region D
Normal Width b	$\theta = 4.75°$	$\theta = 7.5°$
Normal Half-width b_1	$\theta_1 = 2.3°$	$\theta_1 = 3.5°$
Lateral Width l	$\phi = 28°$	$\phi = 60°$
Lateral Half-width l_1	$\phi_1 = 14°$	$\phi_1 = 24°$

The rate of spreading parallel to the wall is seen to be about six to eight times greater than that normal to it. This is due to the effect of the wall in broadening the jet. The increased surface area of the jet becomes available for mixing which results in faster reduction of velocities and greater expansion of the lateral boundaries than in the case of the free jet.

The origin of the distance x in Eq. (2) is theoretically that point from which the jet appears to originate. It is clear, therefore, that if the jet is to be considered as coming from a point source then the location of that point requires to be established experimentally. This distance x_o is found by extrapolating the lateral width line in region C to its virtual origin. The value of x_o thus deduced is about 5 diameters downstream of the orifice. For the case of the free jet, values from 0 to 1.5 diameters downstream of the orifice have been reported for x_o.

Near the wall, the width l_w and half-width l_{w1} expand linearly with divergence angles of 32.5 deg. and 11.5 deg. respectively for both regions C and D of the jet, Fig.10. The corresponding virtual origin for these widths is about three diameters downstream of the orifice.

The development of the boundary layer, extending from the wall to the point of maximum velocity in the profile,

of the flow, the results are plotted as the ratio of velocities divided by the maximum velocity of the particular profile against the ratio of the related position co-ordinates divided by the co-ordinate of the half maximum velocity (the half-width). Representative profiles thus plotted are shown in Figs. 3 to 6.

Concerning the normal velocity profiles in the outer half of the jet (Figs.3 and 4), they vary from section to section in the transition region B and therefore are dissimilar; while the profiles in regions C and D lie very nearly on one curve and therefore are similar despite the different rates of decay of maximum velocity in these two regions. The experimental points are also compared with Tollmien's curve in Fig.4; the agreement between the two results is quite good. Thus the shape of the normal velocity profiles seems to be unaffected by the presence of the wall below.

The lateral velocity profiles along the centreline and near the wall (Figs. 5 and 6) show some variation with distance especially at the outer edge of the jet, where it is also seen that the velocity falls more slowly than in the case of the free jet. It should be noted, however, that part of this discrepancy may be due to inaccuracy of the measurements in this region of the flow.

The normal velocity profiles for the inner half of the jet are plotted non-dimensionally in Fig.7. Each profile consists of a wall layer extending from the surface to the position of maximum velocity and an outer layer comprising a stream of uniform velocity. The profiles are sen to lie conveniently in two groups corresponding to regions C and D already described.

In Fig.8, the profiles for the wall layer in region C are compared with the empirical law of Blasius for the flow in smooth pipes, is shown in Fig.11. The thickness of the layer at first increases to a maximum value of 0.35 d at the end of the transition region B, it remains constant in region C and then decreases gradually in region D to a value of 0.2 d at a distance of 50 diameters. The decrease in thickness in the final region is most probably due to the rapid broadening and transformation of the inner half of the jet into a very wide rectangular stream.

The above trends in the expansion rates of the boundaries, seen also in the decay curves of maximum velocities, seem to indicate a difference in the characteristics of the turbulent mixing in regions C and D of the jet. It is possible that in region D the flow consists of large scale eddies which, because of their size and convective motion, cause the boundary of the jet to oscillate over a larger distance than in region C. It seems such large scale motions bring about a field of smaller eddies uniformly distributed across the flow and so intensify the mixing, and this in turn causes the jet to decay and expand faster.

Notations

The notation used is illustrated in Fig.1 and is given below.

b	Normal width of jet from centreline
b_1	Normal half-width of jet from centreline
d	Jet diameter
l	Lateral width of jet from centreline
l_1	Lateral half-width of jet from centreline
l_w	Lateral width of jet near the wall, y=0.045 in.
l_{w1}	Lateral half-width of jet near the wall, y=0.045 in.
K	Constant defined by equation (1)
u	Mean forward velocity
U_1	Jet exit velocity
U_c	Maximum profile velocity along centreline
U_w	Maximum profile velocity near the wall
x	Longitudinal distance measured from exit section
y	Normal distance measured from the wall
z	Lateral distance measured from centreline
x_o	Distance from point source to the exit section
δ	Boundary layer thickness
θ	Angle of normal boundary divergence
ϕ	Angle of lateral boundary divergence

References

1. J.H. Chesters, C. Holden and A.D. Robertson. "Protection of refractories by moving air curtains. J of Iron and Steel Inst., Vol.85, p.177, Feb. 1957.

2. D. Kuchemann. "Jet diffusion in proximity of a wall. NACA T.M. No.1214, 1949.

3. S.K.A. Naib. "Flow patterns in a submerged liquid jet diffusing under gravity. Nature, Vol.210, p.694, May 14 1966.

4. S.K.A. Naib. "Mixing of a subcritical stream in a rectangular channel expansion". J. Inst. Wat. E., Vol.20, p.199, 1966.

5. S.K.A. Naib. "Photographic method for measuring velocity profiles in a liquid jet". The Engineer, London, Vol.221, p.961, June 24, 1966.

6. S.K.A. Naib. "Unsteadiness of the circulation pattern in a confined liquid jet". Nature, Vol.212, p.753, November 12 1966.

7. G.K. Abramovich. "The theory of turbulent jets". M.I.T. Press, Boston, Mass., U.S.A., 1963.

8. W. Tollmien. "Calculations of turbulent expansion processes". NACA T.M. No.1085, 1945.

9. H.B. Squire. "Jet flow and its effects on aircraft". AIRCRAFT ENGINEERING, Vol.22, p.62, March 1950.

10. A.M. Kuethe. "Investigation of the turbulent mixing regions formed by jets". J. Appl. Mech. Vol.2, p.A 87, 1935.

11. M.B. Glauert. "The Wall Jet". J. Fluid Mech., Vol.1, p.625, 1956.

12. P. Bakke. "An experimental investigation of a wall jet". J. Fluid Mech., Vol.2, p.212, 1957.

13. F.A.L. Winternitz. "Simple Shielded Total-Pressure Probes". AIRCRAFT ENGINEERING, Vol.30, p.313. 1958.

DEFLECTED ROUND JET

Introduction

A jet or a stream of air blown into the atmosphere mixes with the latter and spreads out linearly in accordance with well known laws. In the case of a jet issuing from a nozzle or an orifice and is directed at an angle to a plane surface, it is transformed into a thin stream with wide lateral boundaries at a short distance downstream of the outlet, as shown in Fig.1. The increased surface of the jet becomes available for mixing and energy dissipation which results faster reduction of the velocity and energy of the jet than in the case of the free jet. Between this stream and the surrounding fluid, a turbulent mixing region is formed having normal and lateral boundaries b and l. The normal velocity profile at any section consists of a wall boundary layer and an outer jet layer, which is similar in character to that of the wall jet analyzed by Glauret [1] and the parallel wall jet recently investigated by the author [2].

The behaviour of the deflected jet is of importance in many industrial designs, including combustion chambers, furnaces, jet exhaust systems, stilling basins of pipe chutes and fish passes of the pool type. In the last two cases if the pipe outlet or orifice is inclined downwards, the floor of the pool takes the place of the deflecting plane and this leads to a more rapid velocity reduction than if the outlet or orifice is horizontal. This allows for a shorter length of pool than that required with a horizontal outlet.

Another aspect of the inclined jet design is the steadying effect it has on the flow in the stilling basin of a pipe outlet. It was observed that a round water jet discharging into a rectangular channel produces both unsymmetrical and unsteady flow. For semi-submerged flow, the jet is sucked laterally and adheres to one side of the channel, with a long circulation zone forming on the other side. At higher depth of submergence, the jet changes into a state of sustained oscillatory movement from one side of the channel to another. Attempts to stabilize the flow without altering its character were unsuccessful. However, reasonably steady and symmetrical flow was obtained when the jet was directed at 45° angle to the channel bed.

The data on deflected jets are few and are scattered in a number of publications. Chester et al [3] carried out experiments on air jet from 25mm diameter orifice and

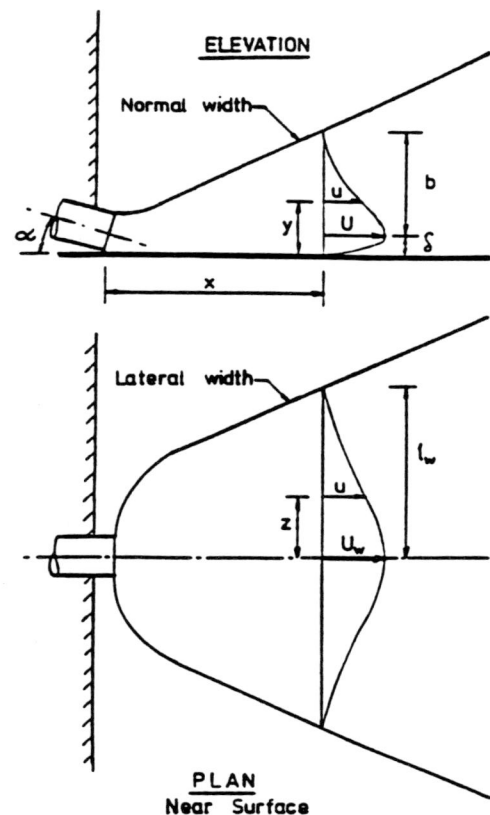

1/ Flow in a deflected round jet.

from a 6.5mm rectangular slot of the same area which were directed over a plane wooden surface. The jets were examined for each aperture with the plane set both parallel to the jet axis and at an impinging angle of 15° to the direction of discharge. For the circular orifice, they found inclination of the jet to the surface causes the angle of boundary divergence across the surface to increase from 50° to 68°.

Nemenyi and White [4] investigated the design of fish passes with submerged orifices given a downward slope of 45°. They found immediately after issue the jet fanned out to 90° on the floor of the pool and ran up the side walls very smoothly, practically the whole of its excess energy being dissipated before it reached the next orifice. This produced a saving in length of pool of about 50 per cent.

Recently, the author has investigated the spreading and development of turbulent jets and streams [5 to 9] including a round air jet projected parallel to a wall [2]. Experiments were carried out to establish the shape of the velocity profiles, the decay of maximum velocity and the rate of growth of the jet. It was found that the jet consists of an inner half extending from the wall to the centre line of the jet where the flow resembles that along a flat plate and an outer half which is similar in character to that of a free air jet discharging into the atmosphere. In the outer half the rate of spreading parallel to the wall is about six to eight times greater than that normal to it. The present research was undertaken to extend this work by studying the flow characteristics of the jet impinging at 15, 30 and 45 degrees to a plane smooth surface.

Experimental method

A jet of air, 25.4mm diameter, was produced by a blower with a rheostat for adjusting the speed of the fan. The jet was directed at different angles to a smooth wooden surface. Except of the small boundary layer on the inside of the nozzle, the velocity profile was practically uniform across the exit section.

A shielded pressure probe [10] with a central sting was used for velocity measurements. The probe has been found to be insensitive to direction changes through a range of angles $\pm$ 45 degrees for a 1 per cent error limit. Near the wall, the velocity profiles across the surface were measured by a total-head tube 1mm inside diameter. Each probe was clamped into a micrometer screw transversing mechanism graduated to 0.05mm.

Analysis of turbulent round jet

The theoretical aspects of the dispersion of a round jet into the stationary atmosphere are covered in numerous papers and only a brief survey is given here to help explaining the trends of the present experimental results.

Tollmien [11] using Prandtl's mixing length theory, solved the fundamental equations of turbulent motion of a round submerged jet originating from a point source. He established the shape of the velocity profiles and showed that the central velocity of the jet U is inversely proportional to the distance x.

$$\frac{U}{U_1} = K \frac{d}{x}$$

Where U_1 is the initial velocity of the jet, K is a constant for the jet, and d is the orifice diameter. The value of K has been found experimentally to be variable depending on the effect of the boundary layer in the nozzle. Based on theoretical and experimental results, Squire [12] recommends a mean value of 6.5.

The analysis also gives the outer boundary of the jet by the following:

$$b = c_1 x$$

the value of c_1 is 0.214, which makes the angle of divergence approximately 12° with the jet axis.

Kuethe [13] extended Tollmien's work to the case of a jet with a finite source and established the length of potential core is about 5 diameters from the orifice. This result agrees with experiments over a wide range of Reynolds numbers.

The flow of a round jet impinging on a plane surface and spreading radially over it has been analyzed by Glauert [1]. He termed the flow as a radial wall jet for which he obtained approximate similar solutions in regions far downstream of the point of impingement.

The analysis near the wall was based on Blasius law for pipe flow, while Prandtl's mixing-length hypothesis was used further out. An experimental investigation of the radial wall jet has been carried out by Bakke [14]. Measurements of velocity profiles were made at distances 10 to 20 times the jet diameter. At these distances, the profiles were similar, and rate of spreading and decay of the jet followed power laws:

$$b \propto x^{1.02}$$
$$U_c \propto x^{-1.11}$$

These are in fair agreement with Glauert's predictions.

Variation of maximum velocity

The variation of maximum velocity of the jet for different angles of deflection are plotted non-dimensionally in Fig.2. As for the parallel wall jet [2] the velocity for the case of 15° deflection is seen to be constant over 2 diameters followed by a transitional length up to 8 diameters which joins to a straight line up to 30 diameters and this is followed by another line of greater slope. With increasing deflection angle the length of transition decreases from 8

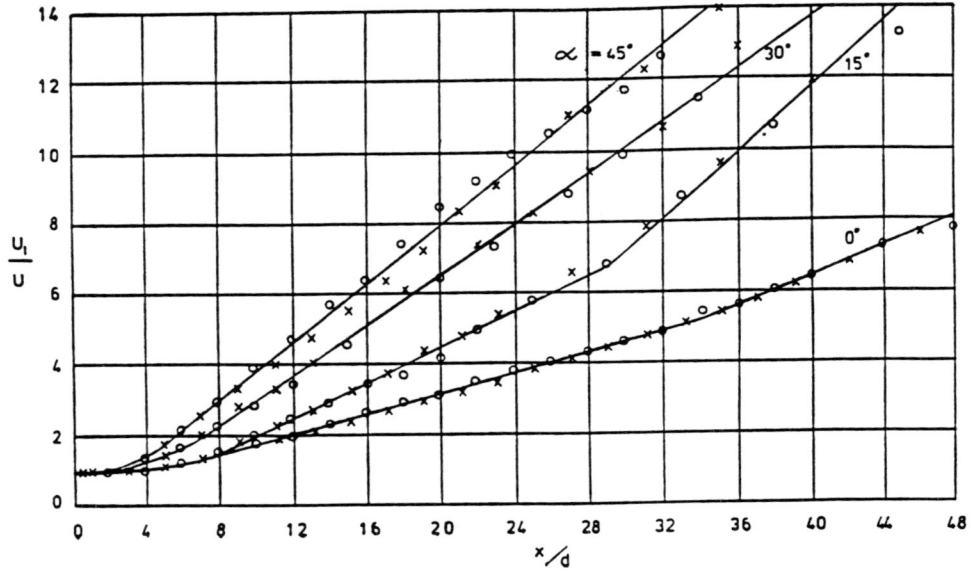

2/ Decay of maximum velocities.

to 6 diameters, and the velocity beyond this length decreases linearly up to 40 diameters.

The length of the potential core is seen to be independent of the deflection angle and it is about 2 diameters compared with 5 diameters for the free round jet. As expected, the presence of the wall on one side of the jet increases lateral mixing and hence reduces the length of the potential core. However, the length of the transition region for the 15° deflection is seen to be such that the linear distribution begins at a distance of 8 diameters as for a free jet. It seem that the impingement region is relatively small and the deflection of the jet is completed within the potential core.

The following relationships between maximum velocity and distance are derived:

Angel	Relationship	Distance
0°	$U/U_1 = 7\,d/(x + 2.5)$	$8 < x/d < 33$
	$U/U_1 = 5\,d/(x - 8)$	$33 < x/d < 53$
15°	$U/U_1 = 4\,d/(x - 2)$	$8 < x/d < 29$
	$U/U_1 = 2.2\,d/(x - 14)$	$29 < x/d < 45$
30°	$U/U_1 = 2.8\,d/(x - 2)$	$7 < x/d < 40$
45°	$U/U_1 = 2.4\,d/(x - 1)$	$6 < x/d < 36$

The origin for each relationship was located by extrapolating the straight line thus obtained to meet the x-axis.

As for the free air jet and the parallel wall jet, four regions of flow are apparent in the 15° deflected jet: (i) Region I of potential core extending about 2 diameters from the orifice; (ii) Region II of transition flow that extends up to 8 diameters and in which the velocity apparently varies parabolically with distance; (iii) Region III of established flow which extends to about 30 diameters and (iv) Region IV of terminal flow in which the residual velocity decays rapidly as a result of large scale turbulence to that of the surrounding air.

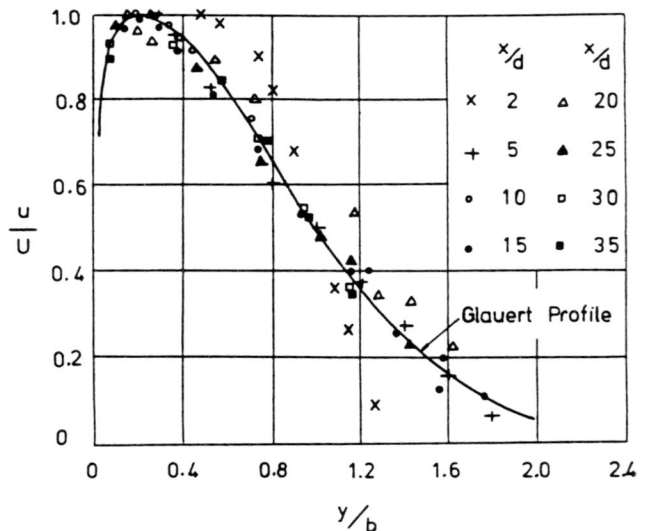

3/ Normal velocity profiles for α = 15°.

36

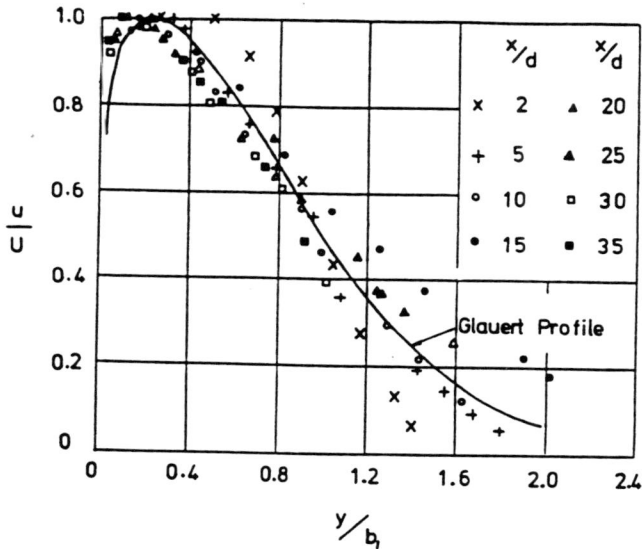

4/ Normal velocity profiles for α = 30°.

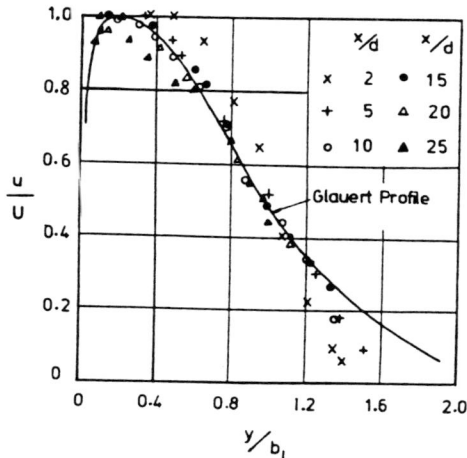

5/ Normal velocity profiles for α = 45°.

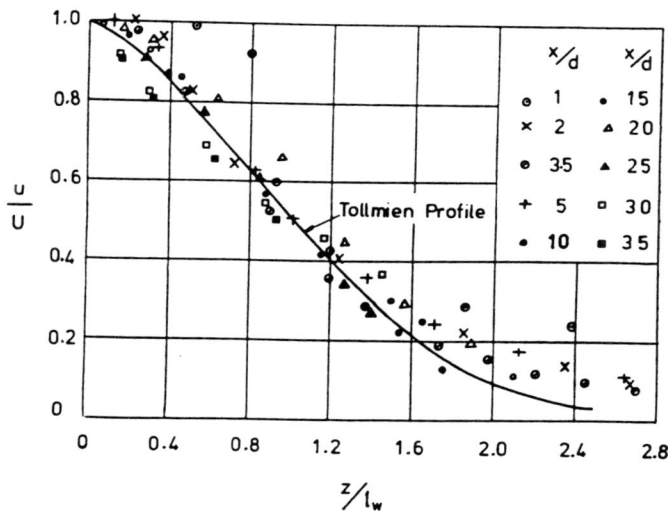

6/ Lateral velocity profiles near the wall for α = 15°.

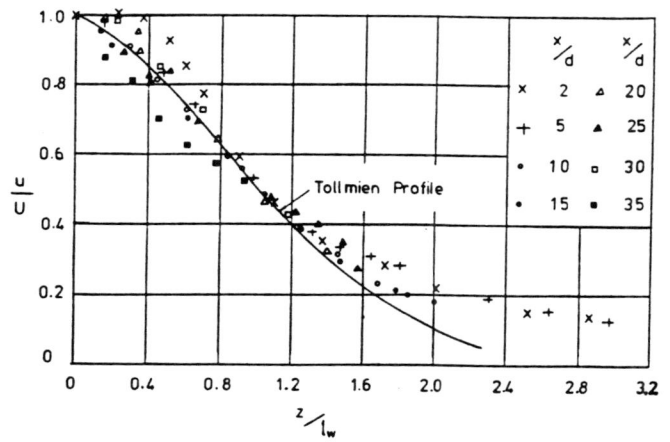

7/ Lateral velocity profiles near the wall for α = 30°.

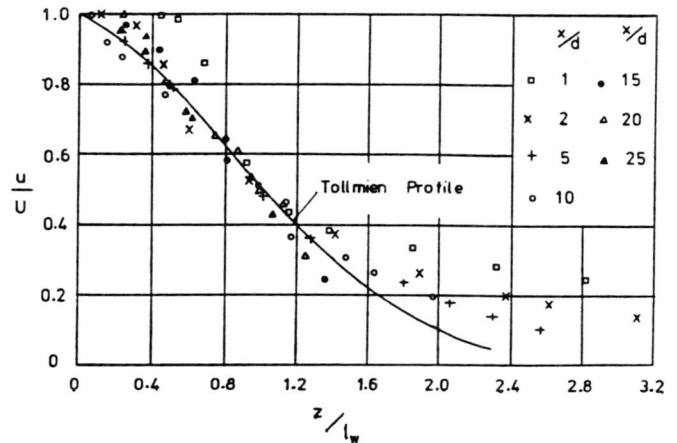

8/ Lateral velocity profiles near the wall for α = 45°.

Because of the greater rate of spreading and hence mixing for the 30° and 45° deflected jets, the rate of decay of maximum velocity is much faster than for the other jets, being about two and a half times that for the parallel wall jet. Also, it appears that downstream of transition Region II, there is only one region of fully developed flow with velocity decay characteristics nearly the same as in Region IV of terminal flow for the 15° deflected jet.

Velocity profiles

The velocity profiles in various regions of the flow are plotted non-dimensionally in Figs. 3 to 8. Concerning the normal profiles along the centre plane of the jet, for regions $5 < x/d < 35$ they lie nearly on one curve and therefore are similar despite the different rates of decay of maximum velocity in the three regions of the jet. The profiles are also seen to be in some agreement with Glauert's profile for the radial wall jet, and thus the two flows are similar.

However, in the outer layer of the jet and near the position of maximum velocity, there is a difference in the velocity profile from that of the wall jet. This is caused primarily by the spreading effect of the wall on the jet and the effect

11/ Lateral width of the jet for α = 45°

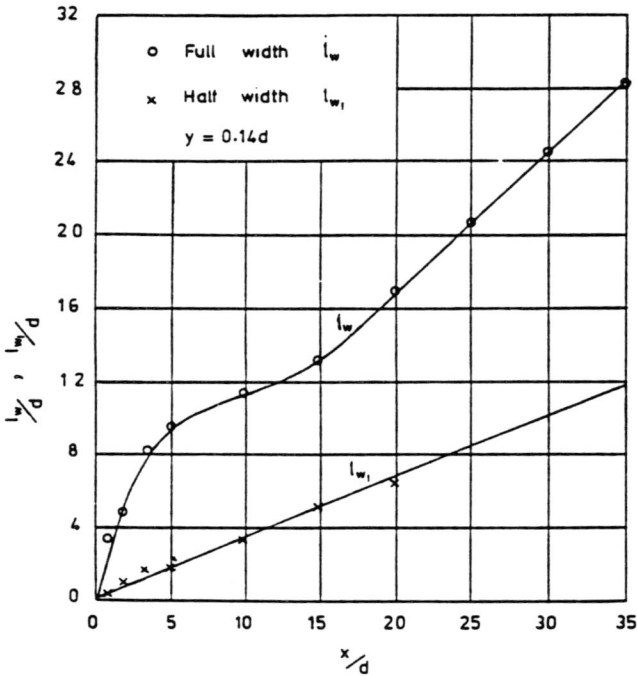

Full width l_w (○), Half width l_{w_1} (×), $y = 0.14d$

9/ Lateral width of the jet for α = 15°.

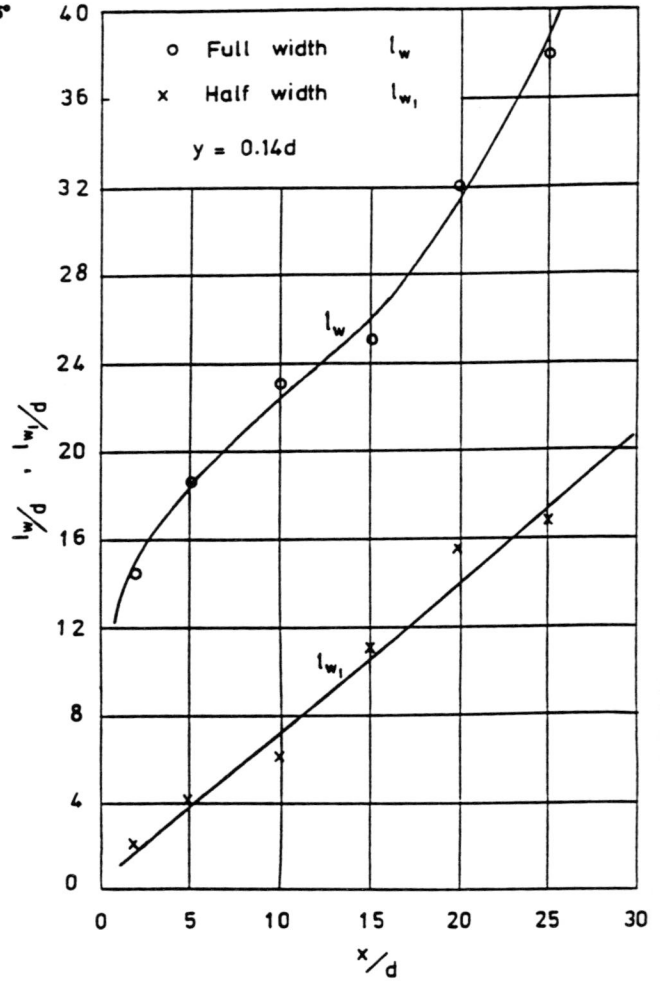

Full width l_w (○), Half width l_{w_1} (×), $y = 0.14d$

10/ Lateral width of the jet for α = 30°.

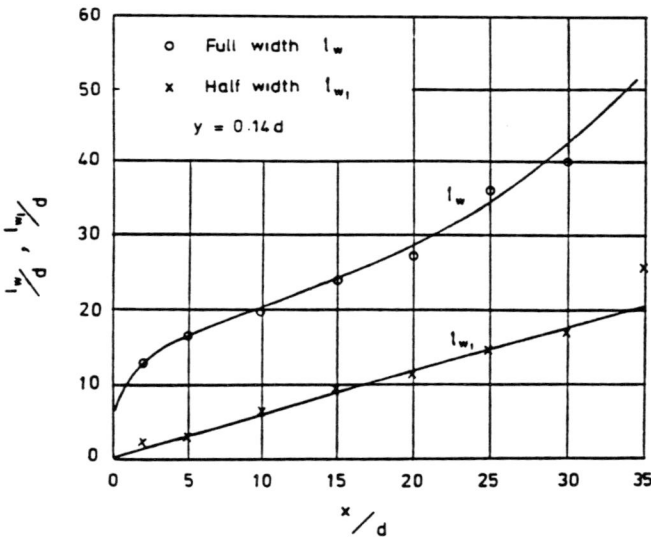

Full width l_w (○), Half width l_{w_1} (×), $y = 0.14d$

12/ Lateral width of the jet for α = 15°.

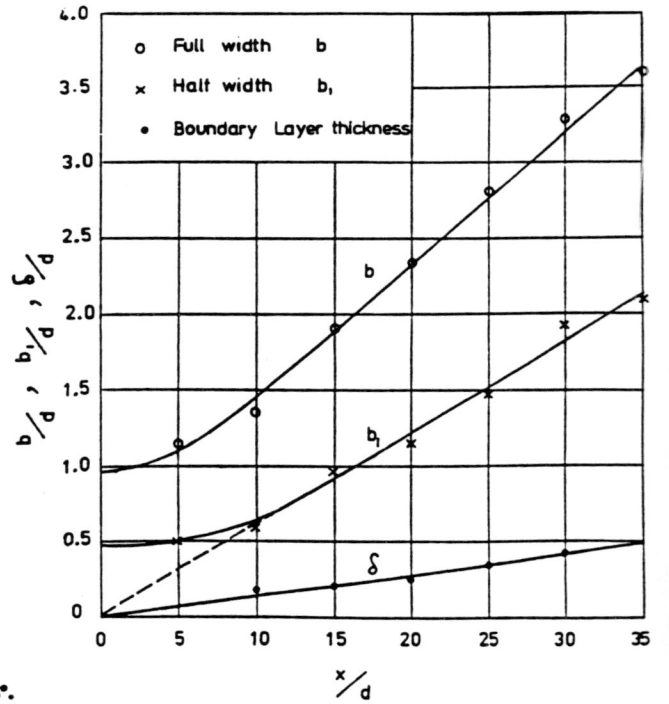

Full width b (○), Half width b_1 (×), Boundary Layer thickness (●)

38

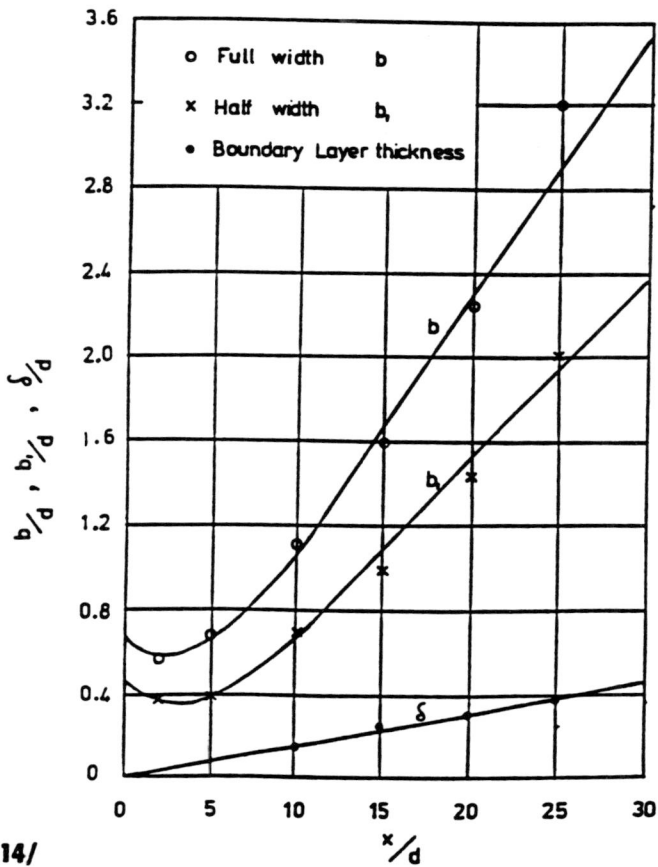

13/

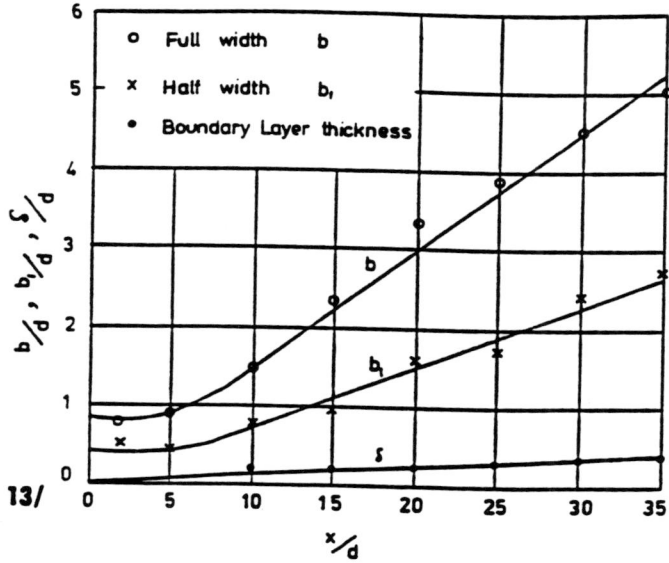

15/

13/ Normal width of the jet for α = 30°.

14/ Normal width of the jet for α = 45°.

15/ Flow in the jet for different angles of impingement.

16/ Comparison of lateral widths of the jets.

17/ Boundary layer growth in the jets.

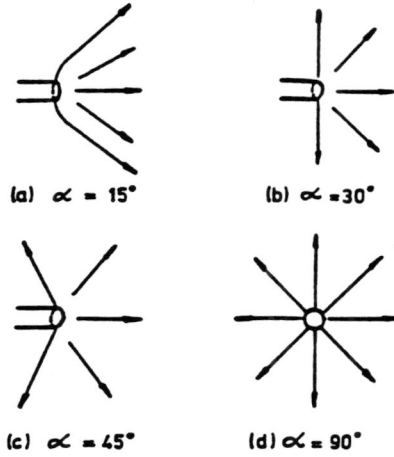

14/

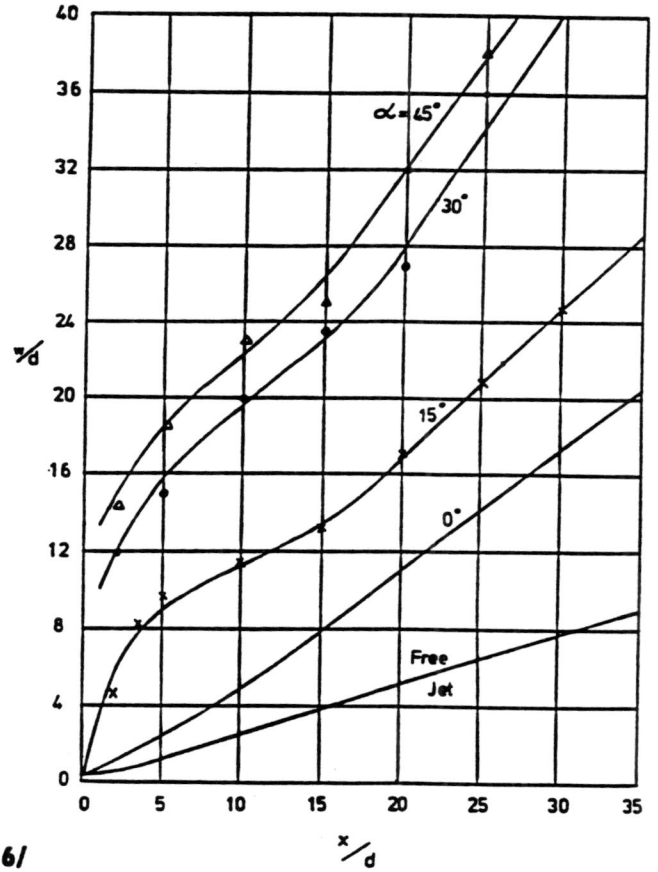

16/

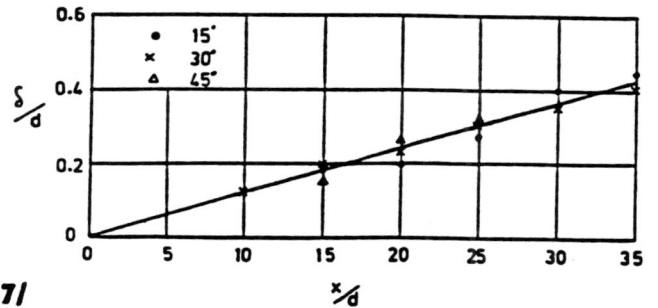

17/

39

of the inner wall layer on the outer layer. The lateral velocity profiles near the wall for $5 < x/d < 35$ show some variation with distance especially at the outer edge of the jet, where it is seen that the velocity falls more slowly than in the case of the free jet. It should be noted, however, that a large part of this discrepancy may be due to inaccuracy of measurements in this region of flow. The profiles compare reasonably with Tollmien's curve for the free jet.

Spreading of boundaries

The normal and lateral widths of the jet for different deflection angles are plotted non-dimensionally in Figs. 9 to 14. On impact the jet spreads out rapidly over the wall, reaching a wide lateral width l_w at a short distance downstream of the nozzle. Using smoke and fine strings as tracers, it was observed that at about 30° deflection angle the jet fans out to 90° on the surface of the wall (Fig. 15). With increasing deflection angle upstream flow occurs (Fig. 15c), until when the jet impinges vertically on the surface, it spreads out radially in all directions (Fig. 15d) as in the case of the radial wall jet. This explains the considerable increase in lateral width of the jet at impact for the 30° and 45° deflection angles and may also account for the similarity of decay of maximum velocity in both cases. Downstream of flow Region II, the lateral width follows a curve which then joins to a straight line of large divergence angle. As expected, the divergence angles for the 30° and 45° deflected jets are the same, about 60°. The widths are compared with that for the round free jet in Fig.16, which shows the considerable increase in lateral width with the increasing deflection angle.

The normal widths, b along the centre plane are plotted in Figs. 12 to 14. It is seen in the impingement Region, the jet contracts slightly and then spreads linearly with maximum slope of about 13° for the 30° defected jet. The angles of normal and lateral boundary divergence Θ and Φ in the regions of fully developed flow are given in the table below.

ANGLES OF DIVERGENCE

Boundary	0° Deflect.	15° Deflect.	30° Deflect.	45° deflect.
Normal width b	4.75°	5.2°	12.7°	7°
Normal half-width b_1	2.3°	2.6°	4.4°	4.9°
Lateral width l_w	28°	37°	59.5°	59.5°
Lateral half-width l_{w1}	14°	17.5°	20°	34°

The rate of spreading parallel to the wall is seen to be about 5 to 7 times that normal to it. This is due to the effect of the wall in broadening the jet on impact and transforming it into a thin wide stream. The increased surface area of the stream becomes available for mixing which results in faster reduction of velocities and greater expansion of the boundaries than in the case of the free jet. The half-widths l_{w1} and b_1 also vary linearly with distance for the jets.

The development of the boundary layer, extending from the wall to the point of maximum velocity in the profile, is shown in Fig. 17. The points lie closely on a straight line passing through the origin, which is represented by the following equation:

$$\frac{\delta}{d} = 0.012 \left(\frac{x}{d} \right)$$

The value of the constant is considerably less than the average value of 0.067 found experimentally for the plane wall jet [15]. This is a further illustration of the deflection effect of the solid surface in transforming the jet into a very wide shallow stream.

Notation

The notation used is illustrated in Fig. 1 and is given below.

b	Normal width of jet.
b_1	Normal half-width of jet.
d	Jet diameter.
l_w	Lateral width of jet near the wall.
l_{w1}	Later half-width of jet near the wall.
K	Constant defined by equation (1).
U_1	Jet exit velocity.
U	Maximum profile velocity.
U_w	Maximum profile velocity near wall.
y	Normal distance measured from the wall.
z	Lateral distance measured from centreline.
x_o	Distance from point source to the exit section.
δ	Boundary layer thickness.
Θ	Angle of normal boundary divergence.
Φ	Angle of lateral boundary divergence.

References

1. Glauert, M.B. "The wall jet". J. Fluid Mech., Vol.1, p.625, 1956.

2. Naib, S.K.A. "Spreading and development of the parallel wall jet". Aircraft Engineering, Lond., p.30, December 1969.

3. Chesters, J.H., Holden, C. and Roberston, A.D. "Protection of Refractories by Moving Air Curtains". J. of Iron and Steel Inst., Vol. 85, p.177, February 1957.

4. Nemenyi, P. and White, C.M. "Report on hydraulic research on fish passes; Appendix of report of the Committee on Fish Passes". Inst. Civil Eng., Lond., 1948.

5. Naib, S.K.A. "Mixing of a subcritical stream in a rectangular channel expansion. J. Inst. Wat. E., Vol.20, No.3, p.199, 1966.

6. Naib, S.K.A. "Stream boundaries of subcritical flow in a trapezoidal channel expansion". "Wat. & Wat. E., Vol.73, pl.155, April 1969.

7. Naib, S.K.A. "Flow patterns in a submerged liquid jet diffusing under gravity". Nature, p.694, May 14, 1966.

8. Naib, S.K.A. "Photographic Method for Measuring Velocity Profiles in a Liquid Jet". The Engineer, Lond., Vol.221, p.961, June 24 1966.

9. Naib, S.K.A. "Unsteadiness of the circulation pattern in a confined liquid jet". Nature, Vol.212, p.753, November 12, 1966.

10. Winternitz, F.A.L. "Single shielded total pressure probes". Aircraft Engineering, Vol.30, p.313, 1958.

11. Tollmien, W. "Calculations of turbulent expansion processes". N.A.C.A., TM No. 1085, 1945.

12. Squire, H.B. "Jet flow and its effects on aircraft". Aircraft Engineering, Vol.22, p.62, March 1950.

13. Kuethe, A.M. "Investigation of the turbulent mixing regions found by jets". J.Appl. Mech., Vol.2, p.A87, 1935.

14. Bakke, P. "An experimental investigation of a wall jet". J. Fluid Mech., Vol.2, p212, 1957.

15. Schwarz, W.H. and Cosart, W.F. "The two-dimensional turbulent wall jet". J. Fluid Mech., Vol.10, Pt.4 p., 481, June 1961.

PART IV

CHANNEL EXPANSIONS AND TRANSITION STRUCTURES

Subcritical Channel Transitions
Rectangular Channel Expansion
Trapezoidal Channel Expansion

SUBCRITICAL CHANNEL TRANSITIONS

Introduction

The design of hydraulic projects often involved changes in shape and cross-sectional area of the stream, whereby the uniformity of the flow is locally disturbed. A structure which is designed to reduce this disturbance and also to conserve the energy of the flow is called a transition. Where the flow is converging its purpose is to minimise energy losses, and where the flow is diverging its purpose is to recover as much as possible of the velocity head. Besides the hydraulic performance, transition design is equally governed by the economic importance of the structure.

Transitions are incorporated in the design of canal flumes, channel contractions and expansions, flumed falls, inlets and outlets for aqueducts, culverts, siphons and tunnels. Transition flumes are often combined in the design of other canal structures with economic advantage. For example, where a channel is to be crossed by a bridge, it is possible to constrict its width by a suitable transition to keep the construction cost of the bridge minimum. Also, in the improvement of natural streams and channels to carry greater discharge, it is more economical to include a constriction transition under an existing bridge than to widen the bridge span.

The main principles of transition design for subcritical flow were first developed by Hinds (1928) and were based on practices of the U.S. Bureau of Reclamation. His paper describes in detail the design of warped transitions for flumes and siphons and gives many examples of transitions suited to different conditions. Scobey (1933) studied the flow of water through flumes, and suggested formulae and coefficients for the calculation of surface profiles and head losses in their inlet and outlet transitions. Hinds' and Scobey's works have been synthesised and reported briefly by Ippen (1950). Montagu (1934)

developed further and demonstrated Hinds' principles by presenting worked design examples of canal flumes.

As a further contribution, the present paper was written to explain the analysis of transitions by the use of the energy diagram, and to present generalised methods of design in a form readily usable by engineers.

Analysis of Flow

The variation of flow through a transition, like other forms of rapidly varied flow, takes place over a relatively short distance, in which acceleration plays the dominant part while boundary friction is comparatively unimportant. An approximate solution of such a flow (Fig.1) may be obtained by assuming the effect of curvature of the streamlines and hence the vertical component of velocity,

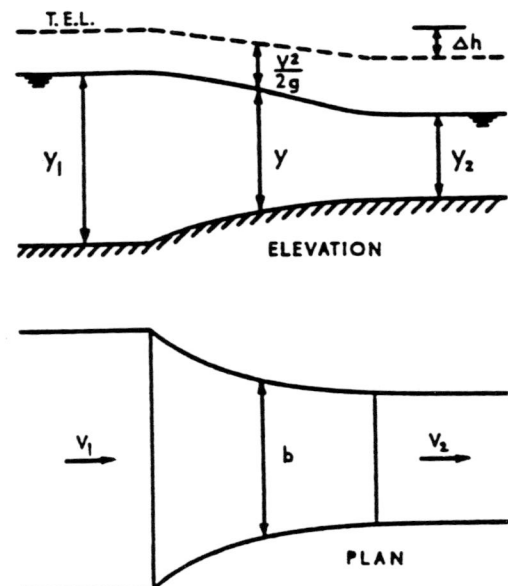

Fig. 1—Subcritical flow in a channel transition

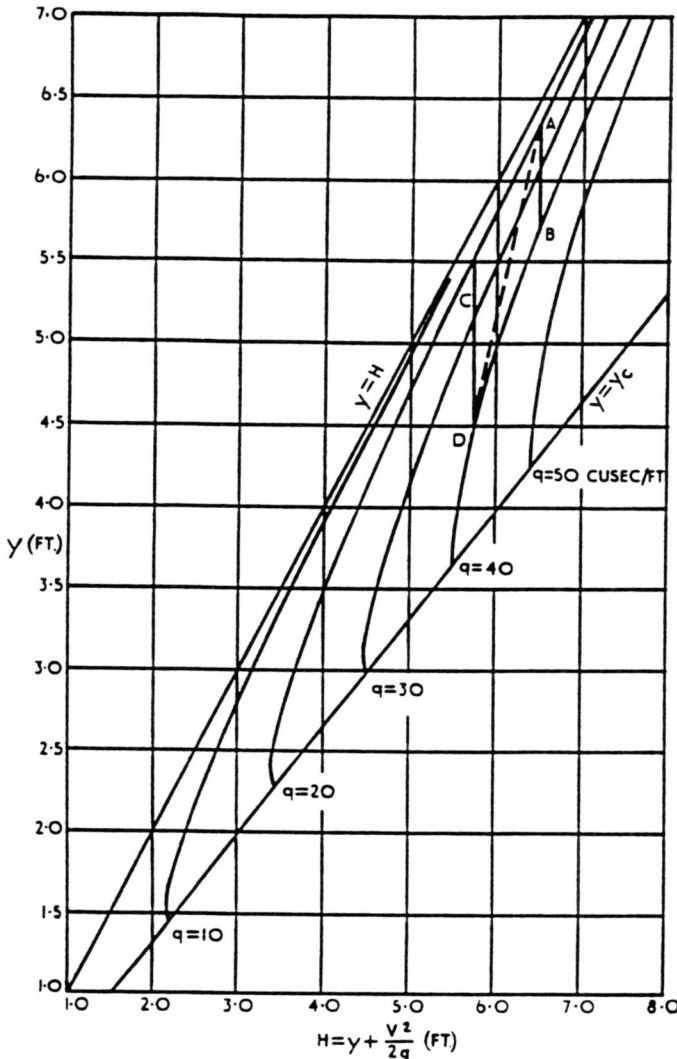

Fig. 2—Application of energy diagram to transition design

The second equation is best solved by a graphical method in which a series of specific energy curves (Fig.2) are prepared for the range of discharges and depths involved in a particular design.

To illustrate the application of the energy curves, consider a transition in which the cross-sectional area is reduced by decreasing the width and raising the bed elevation. With point A in Fig.2 representing the hydraulic conditions at the entrance of the transition, the conditions at the exit represented by point D, may be arrived at by one of the following procedures:

(1) The bed may be continued at the same level up to a section with conditions represented by point B; thus keeping the specific energy H constant while the channel width is gradually reduced. In this operation q is increased while y is decreased, as represented by the vertical line AB. The decreasing values of y are read from the intersections of AB with the curves of q. Next, the channel bottom is gradually raised while keeping the width constant; this process is represented by the line BD.

(2) The two operations above may be reversed: first, the bed is raised while the width is kept constant as represented by the curve AC; then the width is reduced while the bed level is kept constant as represented by the curve CD.

(3) The two operations may be combined into a single process represented by the line AD, with simultaneous adjustments in width and bed level, as shown in Fig.1.

The last procedure is hydraulically the best method of design, as the streamlining of the transition is most gradual and with less likelihood of flow separation. Since it is possible to draw an infinite number of lines between points A and D, the best form of the curve is open to discussion. It depends on the required forms of the sidewalls and bed profile and can not be predicted theoretically for any particular design. Under these conditions and based on past experience, it is reasonable to assume a straight line between the two points as the transition characteristic line.

to be small and negligible; so that the pressure is hydrostatic throughout the flow. Under these conditions, the flow may be analyzed by the continuity and specific energy relationships:

$$Q = Va$$

$$H = y + \frac{V^2}{2g} = y + \frac{Q^2}{2ga^2}$$

where a is the cross-sectional area of flow. For transitions with vertical sidewalls, the flow may be considered two-dimensional with unit discharge q = Q/b and for which the above equations become:

$$q = Vy$$

$$H = y + \frac{q^2}{2gy^2}$$

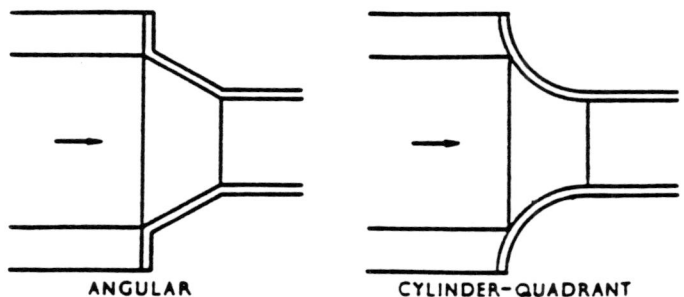

ANGULAR   CYLINDER-QUADRANT

Fig. 3—Low velocity transitions

45

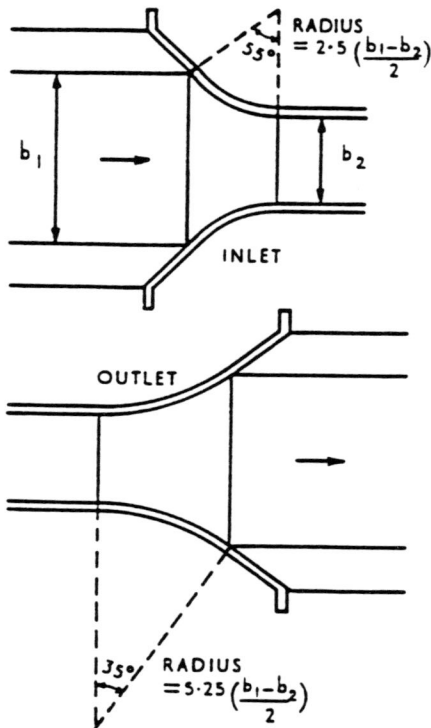

Fig. 4—Curved inlet and outlet transitions

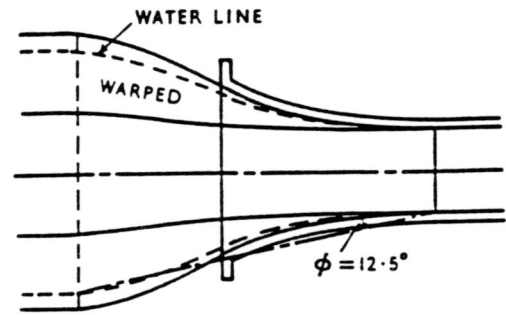

Fig. 5—Warped flume transition

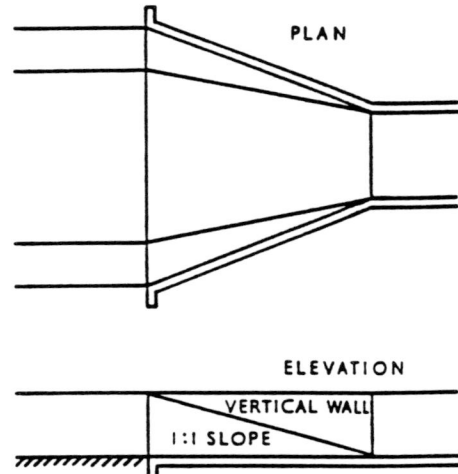

Fig. 6—Wedge-shaped transition

Methods of Design

With the use of the specific energy diagram, three methods of transition design are possible:

(1) The profile of the water surface is sketched in between the levels at the beginning and end of the transition, and then the sidewalls and bed are designed therefrom.

(2) The bed profile is sketched in between the ends of the transition and then the sidewalls are designed therefrom.

(3) The sidewalls are sketched in between the ends of the transition and then the bed is design therefrom.

In the first method, a trial-and-error procedure is necessary for finding the values of y and H from the energy diagram knowing only the velocity head; this makes the method rather tedious and unpracticable. In the second method, the bed profile is drawn arbitrarily; therefore the estimation of head loss for this method is inaccurate. For the third method, a number of standard sidewalls have been developed for which the energy losses can be calculated accurately. For this reason, the last method is most convenient for practical design. However, in the design of partially-closed transitions, such as the inlet and outlet transitions of inverted siphons, both the sidewalls and bed profile must be sketched in, and then the water surface and position of the siphon roof are determined.

Types of Transitions

The form of a transition may vary from the simple type with straight sidewalls to very elaborate streamlined structures. Figs.3 to 6 show five types of transition from a trapezoidal to a rectangular channel: (1) straight-sidewall transition set 30 to 45 degrees to the channel axis, (2) cylinder-quadrant transition, (3) curved transition, (4) warped transition, and (5) wedge-shaped transition. The cylinder-quadrant transition is essentially a pair of circular wingwalls tangent to the sides of the rectangular channel and extending into the trapezoidal earth channel as cut-off walls. While very successful as an inlet, this form of outlet is not much improvement over the 45 degrees wingwall and is not recommended for this purpose. Outlet wingwalls set 30 degrees to the channel axis have proved more efficient (Scobey, 1933).

The use of the first two types of transitions should be limited to small unimportant structures and where the velocity is low, not exceeding 4ft/sec. For larger structures, where velocities range from 4 to 8 ft/sec, curved transitions with angular cut-off walls are recommended by the author. Due to the inherent instability of diverging streams, the outlet transition (Fig. 4b) is made to expand over a longer length to prevent flow separation.

46

For important structures, especially where velocities are in excess of 8ft/sec and the Froude number $F = V/\sqrt{gy}$ is greater than 0.5, the warped transition should be used. Here, the sidewalls are sketched in such that a straight line joining the flow line at the two ends of the transition will make an angle of about 12.5 degrees or 1: 4.5 with the axis of the channel. Examples of warped transitions between a canal and a flume, and between a canal and an inverted siphon are described by Hinds (1928); the same are also reported by Chow (1959). The wedge-shaped transition is a simplified form of the warped type, where the warped sidewall is replaced by a vertical and a sloping wall with a straight line intersection running from the top edge of the transition entrance to the bed at the exit.

It is interesting to note that the length of a warped transition may be reduced by the use of intermediate splitter walls with included angles of 7 degrees to guide the flow (Ippen, 1950).

Head Losses

The head loss in a transition is due to turbulence and friction losses, and it is commonly expressed as a coefficient K multiplied by the change in velocity head Δh_v between the ends of the transition:

$$\Delta h = K \, \Delta \, h_v = K \left(\frac{V_2^2 - V_1^2}{2g} \right)$$

In an inlet transition, the velocity of flow is gradually increased; this results in a drop in water level given by:

$$\Delta \, y_i = y_1 - y_2 = \frac{V_2^2}{2g} - \frac{V_1^2}{2g} + \Delta h$$

$$= (1 - K_i) \left(\frac{V_2^2 - V_1^2}{2g} \right)$$

In an outlet transition, the velocity is reduced, and the decrease in velocity head at the end of the transition is partly dissipated and partly converted into pressure head to raise the water level by:

$$\Delta \, y_o = \Delta \, h_v - \Delta h = (1 - K_o) \left(\frac{V_1^2 - V_2^2}{2g} \right)$$

Appropriate values of the head loss coefficients K_i and K_o are shown below:

Type of Transition	K_i	K_o
Angular	0.30	0.35-0.50
Cylinder-quadrant	0.20	0.40
Curved	0.20	0.35
Wedge-shaped	0.15	0.30
Warped	0.10	0.20

The head loss Δh may be assumed to be uniformly distributed along the length of the transition, thus giving the elevation of the total energy line at any section of the structure.

The head losses in sudden expansions and contractions with straight bed profiles have been investigated experimentally by Formica (1955) and reported by Chow (1959). Recently, the author (1966) has investigated the distribution of velocity and rates of spreading of a subcritical stream in a sudden channel expansion.

Sidewall Sketching Method

Given the discharge with corresponding values of depths and areas for the upstream and downstream channels, the procedure of design if summarised below:

(1)   Obtain or prepare a specific energy diagram for the range of unit discharges and depths through the transition.

(2)   Draw suitable sidewalls to reconcile between economy of construction and best hydraulic performance for the particular design.

(3)   Mark a number of sections on the plan and elevation of the structure and measure the successive bed widths. Calculate the unit discharge q for each section and tabulate the results.

(4)   Estimate the head loss Δh in the transition using equation (5), and plot the variation of the total energy along the structure. If the energy loss is neglected, this line will be horizontal.

(5)   Draw the transition characteristic line on the energy diagram, joining the characteristics of the stream at the entrance and exit of the transition.

(6)   Determine the specific energy and depth of flow appropriate to each section from the intersection of the line of transition with the appropriate curve of unit discharge. For intermediate values, the results must be interpolated.

(7)   Determine the bed level at each section by subtracting the value of the specific energy from the level of the total energy line. Draw a smooth mean curve through the points to obtain the bed profile.

(8)   Determine the water surface level by adding the value of the depth to the level of the bed at the section.

If the resulting bed and surface profiles are irregular, they indicate incorrect assumptions of the sidewalls, which should be redrawn and the above procedure be repeated.

Design of Nonprismatic Transitions

In warped and wedge-sharped transitions, the cross-sectional area of the flow is a complex function of the depth and width of the local section; consequently the energy diagram shown in Fig.2 can not be used for design. Under these conditions the following procedure may be used:

(1)   Sketch in the sidewalls and a smooth water surface between the ends of the transition. For warped transitions, the water profile may be assumed as two equal parabolas, tangent to each other at mid-section and horizontal respectively at the two ends.

(2)   Mark a number of sections on the plan and elevation of the structure and measure the successive surface and bed widths s and b respectively. Calculate the mean width at each section $b_m = 0.5 (s + b)$ and tabulate the results.

(3)   Estimate the head loss Δh and plot the variation of the total energy along the transition.

(4)   Measure the velocity head $V^2/2g$ at each section by subtracting the level of water surface from the level of the total energy line, and hence compute the mean velocity V.

(5)   For the known discharge Q, calculate the cross selectional area a = Q/V.

(6)   Calculate the depth of flow at each section $y = a/b_m$.

(7)   Determine the elevation of the bed by subtracting the depth from the level of the water surface.

If the resulting bed profile is irregular, the sidewalls and water surface, should be modified and the above process be repeated.

References

1.   Chow, V.T., 1959. "Open Channel Hydraulics", McGraw-Hill Book Co., N.Y.

2.   Formica, G., 1955, "Experienze Preliminari Sulle Predite di Carico nei Canali, Dovute a Cambriamenti di Sezione", L'Energia ellettrica, Milano, Vol.32, No.7, p.554.

3.   Hinds, J., 1928. "The Hydraulic Design of Flume and Siphon Transitions", Trans, A.S.C.E., Vol.92, p.1423.

4.   Ippen, A.T., 1950, "Channel Transitions and Control", Chapter VIII in `"Engineering Hydraulics", H. Rouse (Editor), John Wiley & Sons Inc., N.Y.

5.   Montagu, A.M.R., 1934, "Fluming", Central Board of Irrigation Publ, No.6, India.

6.   Naib, S.K.A., 1966, "Mixing of a Subcritical Stream in a Rectangular Channel Expansion", J. Inst. Wat. E., Vol.20, No.3, p.199.

7.   Scobey, F.C., 1933, "The Flow of Water in Flumes", U.S. Dept. of Agriculture Tech. Bulletin No. 393.

RECTANGULAR CHANNEL EXPANSION

Synopsis

The sudden expansion of a subcritical stream into a rectangular channel is investigated. By using a shielded total-head probe, the velocity distributions in the mixing region are measured for a Froude number $F_1 = 0.5$ and an expansion ratio $B_2/B_1 = 2$. A brief account is given of the theory worked out by Abramovich for confined flow in the circulation region behind a two-dimensional bluff body. The experimental results are compared with this theory and also the mixing-length theory developed by Tollmien for free jets. In general the agreement between the experimental and theoretical results is satisfactory.

Introduction

When a channel section is suddenly enlarged, the incoming stream gradually spreads out until it occupies the whole width of the larger channel a certain distance downstream. Fig.1 shows a diagram of half the flow. At the junction between the two channels the stream has a constant velocity U_1.

Between this stream and the adjacent slow-moving fluid a turbulent mixing layer is formed having an inner boundary 01 and an outer boundary 02. The line 03 defines the mean flow line with the stream function $\psi = 0$, bounding the fluid which is sucked outward from the stream. The velocity profile at any section XX' consists of a region of constant velocity U_1, a transitional boundary layer, and a constant counterflowing velocity U_2.

The mixing process between the stream and the surrounding fluid causes part of the latter to be carried forward with the stream under conditions in which not only the forward momentum but also total discharge is conserved. The entrainment leads to recirculation to replace the fluid entrained by the stream. Thus, a zone of circulation is established, in which the reverse flow increases with distance from the junction JJ', until it reaches its maximum value at section MM', and then decreases to zero at the stagnation point.

Due to the turbulent mixing process, the circulation

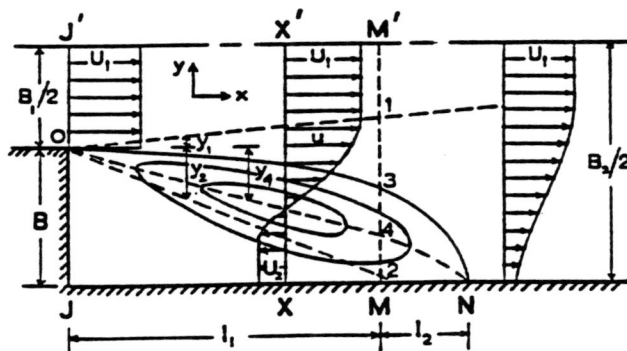

Fig. I
Flow in a channel expansion

zone is a source of energy loss, and past experiments deal mainly with this aspect of the flow. Formica (1955)* carried out tests on gradual and sudden expansions at different discharges from which he obtained average values of head loss coefficients; his results are also reported by Chow (1959). The purpose of the present investigation was to establish the distribution of velocity and rates of spreading of a stream expanding into a rectangular and a trapezoidal channel, on all of which information is required for a proper design of stilling and protection works downstream of control and transition structures. The results for the rectangular expansion are presented here; those for the trapezoidal expansion will be reported in another paper.

Theoretical Analysis

If the pressure is assumed to be hydrostatic and uniform throughout the expansion, then the flow may be analyzed by Abramovich's method (1963) for flow behind a two-dimensional bluff body. In this method, the circulation zone is divided into two regions, MM' being the dividing section. To simplify the problem the effect of wall friction is neglected. In the first region, the velocity profiles in the mixing boundary layer are assumed to be similar and are given by Schlichting's formula (1930),

$$\frac{U_1 - u}{U_1 - U_2} = (1 - \eta^{1.5})^2$$

where $\eta = (y - y_2)/(y_1 - y_2)$. The growth of the boundary layer width is assumed linear, the same as when a free air jet discharges into the atmosphere:

$$b = (y_1 - y_2) = cx = 0.3 \ x$$

The value of the coefficient c is found by experiments. In order to find the rates of spreading and the variation of the reverse velocity, the above relationships are employed in applying the momentum and continuity equations to the control volume of fluid JJ' MM'. The results obtained are shown later in Figs. 6 and 7, the inner boundary 01 is given by:

$$\frac{y_1}{b} = 0.416 + 0.134 \ m$$

where $m = U_2/U_1$. The outer boundary 02 is then

$$\frac{y_2}{b} = \frac{y_1}{b} - 1 = -0.584 + 0.134 \ m$$

For finding the length of the first part of the circulation zone l_1 it is assumed that the amount of energy in the forward and back flows of the circulation zone in section MM' is equal, and in view of the constant pressure in this section, the equality of the total energies is reduced to the equality of the kinetic energies. This hypothesis gives

$$l_1/B = 5.1 \ and \ m_{max} = -0.4$$

where $B = (B_2 - B_1)/2$. This theory is correct as long as in the stream a core of constant velocity (potential core) is retained. Calculations show that the lengths of the potential core and the first region coincide at a relative width of the zone $\beta_c = 0.635$, where $\beta = 2B/B_2$. Obviously, as β exceeds β_c when the potential core has decayed within the limits of the first region of the circulation zone, the above theory becomes less accurate.

The assumption that at section MM' the average kinetic energies in the forward and reverse flow, and also their cross-sections, are equal may justify an approximate analysis of the second region of the circulation zone 3MN by the hydrodynamic method of ideal fluids. The principle of this method is to find first the complex potential of the flow in the hydrograph plane and then by conformal transformation, the streamlines and the geometrical characteristics of the region in the physical plane are determined. Such calculations show that over the range $\beta = 1.35$ to 4.6, the length of the second region of the circulation zone is

$$l_2/B = 0.980 \ to \ 0.852$$

Hence, the total length of the zone is

$$\frac{l}{B} = \frac{l_1 + l_2}{B} = 6.18 \ to \ 6.05$$

or approximately it is equal to

$$l = 6.1 \ B$$

The results for variation in velocity of the reverse flow in the second region is approximated in Fig.6.

Experimental Method

The experiments were carried out in 1ft. wide glass flume having an outlet weir for adjusting the depth of flow and a weighing arrangement for measuring the discharge. The flume was fitted with Perspex guide walls to produce a stream 3 in. deep by 6 in. wide with a Froude number F_1 = 0.5, expanding abruptly on one side into the full width of the flume.

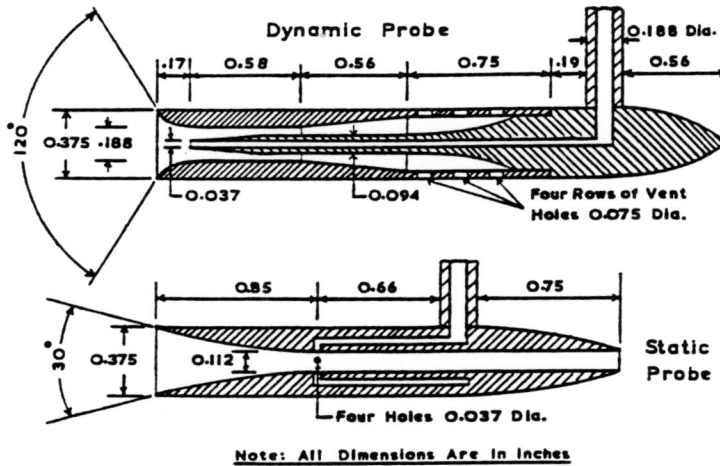

Fig. 2
Shielded dynamic and static probes

Initially, three instruments were designed for measurements: a yaw meter having an angle of 70° (Bryer, Walshe, and Garner 1958), a shielded static probe and a shielded dynamic probe (Russell, Gracey, Letko, and Fournier 1951), Fig.2. On calibration in a free stream, the last two probes were found to be insensitive to directional changes through a range of angles of ± 40°. However, on use in the mixing region, the yaw meter gave bigger angles for the flow direction than indicated by a fine thread, while the static tube gave higher pressures than the local water level measured by a point gauge; consequently, both instruments were disregarded. A Prandtl pitot-static tube was also tried but was found to be subjected to suction.

Generally, the distribution velocity in an open channel is not uniform, the maximum value usually occurs between 0.5 and 0.25 of the depth d below the free surface, while the filament of mean velocity lies between 0.55 and 0.65 d. To show the effect of this type of variation on the diffusion of the stream, lateral velocity profiles were measured at levels Z = 0.2d and 0.6d below the free surface.

Experimental Results

Fig.3, shows the velocity profiles measured at a level Z = 0.2d. To compare their shapes, the profiles are plotted non-dimensionally in Fig.4, together with the three-halves power law of Schlichting.

It is seen that for x/B <3, the profiles vary from section to section and are, therefore, dissimilar; the actual velocity falls more slowly on the steam edge but faster on the circulation side than is assumed in the theory. Further downstream, the profiles tend to settle down to a steady shape, closely following that of equation (1). The difference between the actual and assumed velocity profiles in the first half of the circulation zone is to be

expected, since the assumed distribution was deduced by Schlichting for wake flow at greater distances from the body.

It is interesting to note that experiments by Liepmann and Laufer (1947) on a half jet, consisting of an air stream diffusing into the atmosphere on one side, indicated similar trends: the mixing boundary layer was found to be laminar up to x = 6cm, and fully developed turbulent profiles were established only beyond x = 30cm. The distributions of velocity at a level Z = 0.6d are shown in Fig.5, up to x/B = 3 the profiles are nearly the same as those at level Z = 0.2d, the distortions in the profiles beyond this section are most likely to be due to the formation of secondary spiral vortices in the direction of mean flow. Fig.6, shows the variation in velocity of the backflowing stream plotted on the theoretical curve. Two results are in fair agreement, though a small difference is noticeable; but this may be due to the displacement effect of the boundary layer on the wall JN of the larger channel, which reduces the section of the reverse flow, thereby increasing its velocity.

The curves in Fig.7, show the spreading of the stream boundaries in the channel expansion. The inner boundary is seen to expand at an angle of 5° or about 1 in 12, which is less than the predicted rate for 01, but it is the same, not surprisingly as that for a half-jet analyzed by Tollmien (1926) using Prandtl's mixing-length theory. The mean spreading angle of the outer boundary is about 13°; this is greater than the predicted rate for 02, but closely conforms to Tollmien's solution for a plane free air jet, with an expansion angle of 14° or about 1 in 4. The overall width of the mixing region b is found, however, to be in accordance with equation (2). The disagreement between experiments and Abramovich's analysis is most likely due to the inaccurate assumption of the velocity shape, and also to the errors involved in ascertaining where the mixing ends in the circulation zone.

51

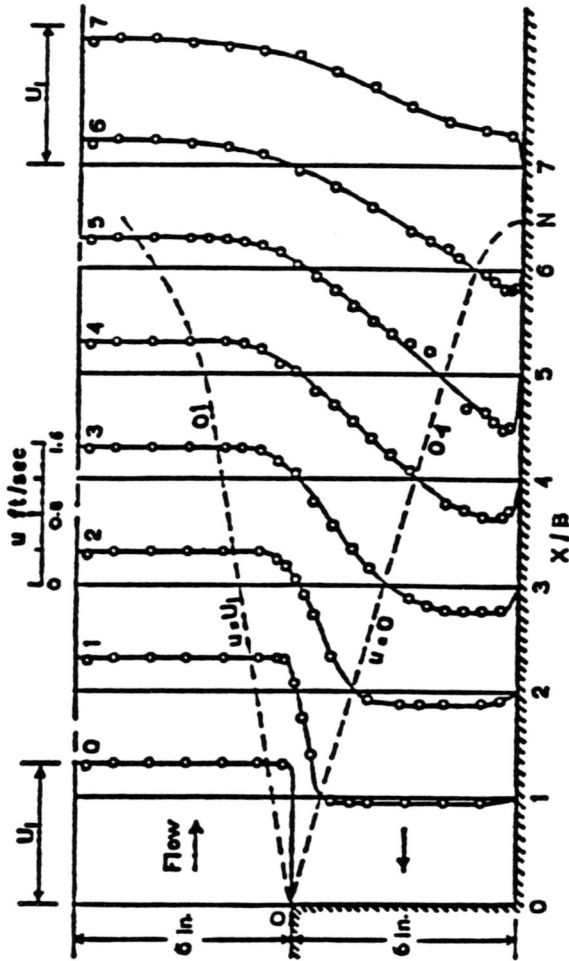

Fig. 4

Fig. 3 (top left)
Velocity profiles at Z = 0·2d

Fig. 4 (above)
Comparison of velocity profiles

Fig. 5 (left)
Velocity profiles at Z = 0·6d

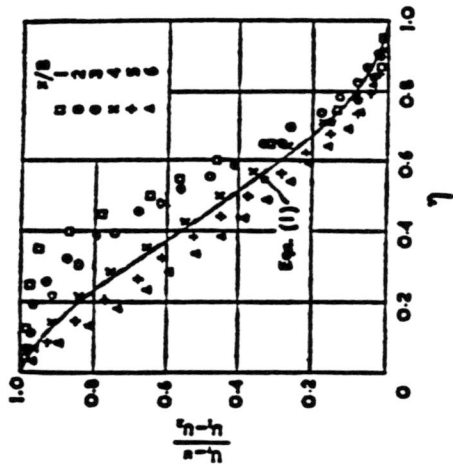

Fig. 3

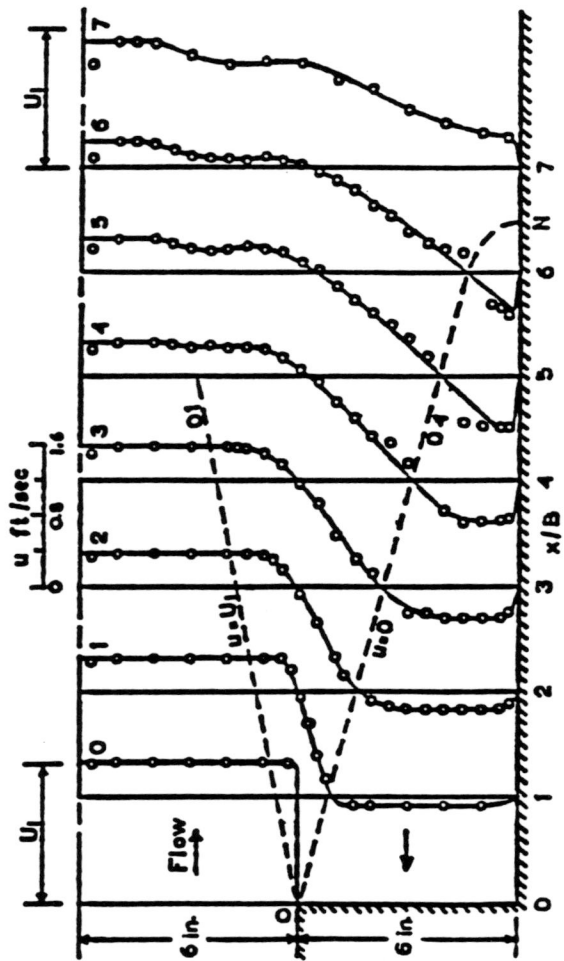

Fig. 5

In the case of an air jet discharging freely into the atmosphere, the entrainment process continues at all distances downstream. Consequently, the velocity profile retains its shape while the jet is spreading continually. In the channel expansion, where there is no external supply of fluid to replace the amount entrained, fluid must be recirculated from the stream into the circulation zone. Apparently, due to this recirculation, and also due to the limited mixing width as the outer boundary 02 reaches the wall JN, the mixing process is intensified, which results in a rapid broadening of the stream until it occupies the whole

width of the larger channel at the stagnation point, which defines the end of the circulation zone. The length of this zone is about 6.5B, compared with the theoretical value of 6.1B.

A calculation of the magnitudes of pressure differences in a free jet was first calculated by Tollmien. He reasoned that since in the central part of the jet the direction of mean velocity is outward, whereas in the outer region of the jet there occurs an inward flow, the static pressure in the latter region must be minimum. Tollmien integrated the appropriate equation of motion and obtained the following relation for the pressure difference between the middle and the boundary of the jet

$$p_o - p_1 = 0.00584 \frac{\rho U_1^2}{2}$$

whereas to create an inflow velocity $V = 0.032\ U_1$ for fluid at rest would require a pressure difference,

$$p_2 - p_1 = 0.00102 \frac{\rho U_1^2}{2}$$

so that the difference from infinite to the centre would be

$$p_o - p_2 = 0.00482 \frac{\rho U_1^2}{2}$$

$$or \quad (p_o - p_2) / \frac{1}{2} \rho U_1^2 = 0.5\ per\ cent$$

where ρ is the density of the fluid. The peak values actually obtained at level $Z = 0.2d$ were around -2 percent. The difference may be mainly due to the presence of large eddies along the outer boundary of the stream, whose centres must contain regions of high negative pressures which have not been included in the theoretical analysis.

The longitudinal variation of static pressure along the wall JN is plotted non-dimensionally in Fig.8; p is the pressure at any point on the wall and p_j is the pressure at the junction section used as a reference. As the fluid in the counter flowing stream is accelerated in the upstream direction, the relative pressure drops from its value of +14 per cent of the dynamic pressure at the stagnation point to

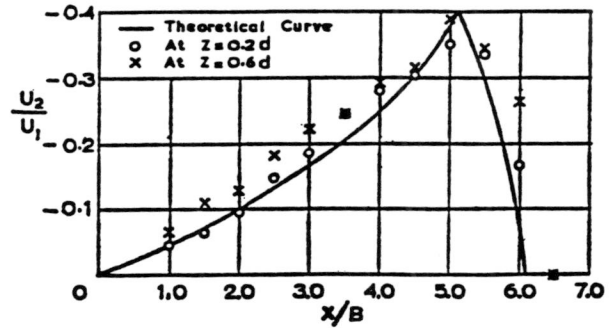

Fig. 6
Variation in velocity of reverse stream

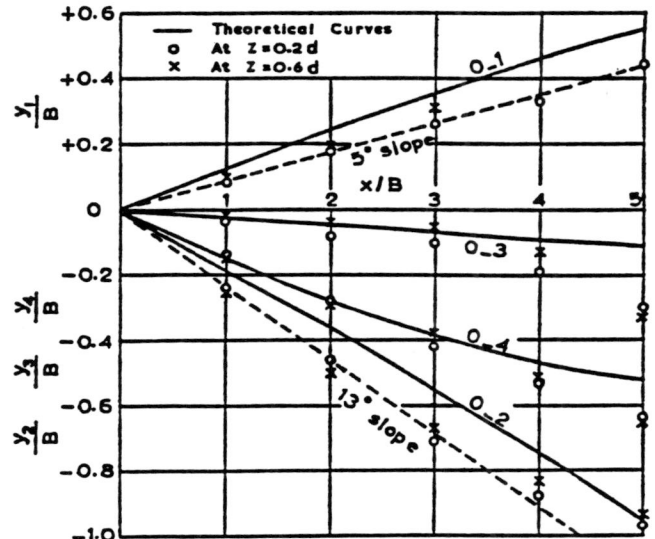

Fig. 7
Spreading of stream boundaries

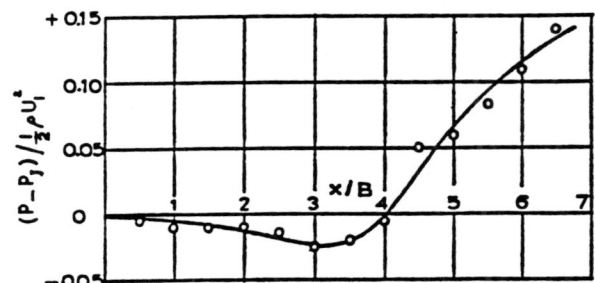

Fig. 8
Variation of static pressure along wall JN

53

a minimum value of -2.5 per cent nearly midway in the circulation zone. However, over most of this zone, the variation in static pressure is much less than 10 per cent; thus the assumption of constant pressure in the theory is reasonably justified.

In the analysis of a half-jet, Tollmien found that along the streamline $\psi = 0$ defining the boundary 03, the velocity is $0.566\ U_1$. Since this streamline terminates at the stagnation point, the velocity head $0.32\ U_1^2/2g$ should, theoretically, be converted into pressure head. From Fig.8, the overall pressure rise is 16.5 per cent of the dynamic pressure. This is only 52 per cent, of the above head; the rest of the energy is probably dissipated by turbulence.

References

1. Abramovich, G.K. 1963, "Theory of Turbulent Jets". M.I.T. Press, Boston, Mass, U.S.A.

2. Bryer, D.W., Walshe, D.E. and Garner, H.C. 1958, "Aeronautical Research Council, R and M. No.3037, "Pressure Probes Selected for Three-dimensional Flow Measurement".

3. Chow, Ven Te, 1959, "Open Channel Hydraulics", McGraw-Hill Book Co., New York and London.

4. Formica, G. 1955, "L'Energia ellettrica, Milano, Vol.32, No.7, p.554, "Experienze Preliminari Sulle Predite di Carico nei Canali, Dovute a Cambiamenti di Sezione".

5. Liepmann, H.W., and Laufer, J. 1947, "National Advisory Council for Aeronautics Tech. Note No. 1247, "Investigation of Free Turbulent Mixing".

6. Russell, W.R., Gracey, W., Letko, W., and Fournier, P.G. 1951, National Advisory Council for Aeronautics Tech. Note No.253, "Wind Tunnel Investigation of Six Shielded Total-pressure Tubes at High Angles of Attack".

7. Schlichting, H. 1930, Ingr. Arch, Vol.5, p.533, "Uber das Ebene Windschatten-problem".

8. Tollmien, W. 1926. Zeitschrift fur ang ewandte Mathematik un Mechanik, vol.IV,p.468, "B e r e c h n u n g   d e r   T u r b u l e n t e n Ausbreitungsvorgange" (Also translated at National Advisory Council for Aeronautics Tech. Note No.789, 1936).

TRAPEZOIDAL CHANNEL EXPANSION

Introduction

For some time, the author has been investigating the diffusion of confined liquid jets and streams, and in particular, attention has been devoted to the flow pattern in their circulation zones[1-6]. In this connection the spreading of a subcritical stream in a rectangular channel expansion proved of special interest, since a theoretical solution for the flow was obtained based on Abramovich's analysis of the flow behind a two-dimensional bluff body[4]. The way the experimental results agreed with theory proved very encouraging and it was decided to extend the work to the case of a stream expanding abruptly into a trapezoidal channel for which no measurements could be traced in the literature.

At any depth in the trapezoidal expansion, the pattern of flow is similar to that of the rectangular case shown in Fig.1. Between the uniform stream with velocity U_1 and the adjacent slow-moving fluid a turbulent mixing layer is formed having an inner boundary 01 and an outer boundary 02. The line 03 defines the mean flow line with the stream function $\psi = 0$ bounding the fluid which is sucked away outward from the stream. The mixing process between the stream and the surrounding fluid causes part of the latter to be carried forward with the stream. The entrainment leads to recirculation to replace the fluid entrained by the stream. Thus a zone of circulation is established in which the reverse velocity U_2 increases with distance from the junction JJ', until it reaches its maximum value at section MM' and then decreases to zero at the stagnation point.

The nature of the present research is essentially the same as that for the rectangular expansion reported in a previous paper; namely, to establish the distribution of velocity, the rates of spreading of stream boundaries and the length of the circulation zone in

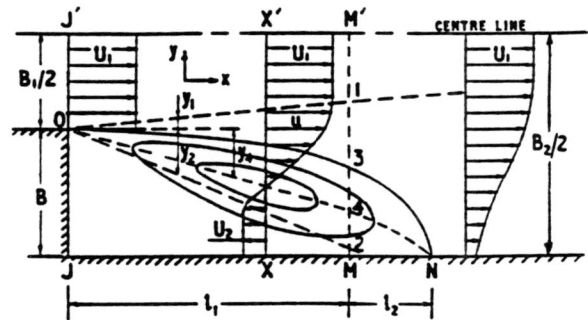

Fig. 1—Flow in a channel expansion

the trapezoidal expansion. Using these results and experimental data for the lift and drag of blocks on bed of a stream, a generalised design is developed for the protection downstream of sudden transitions and energy dissipators[7/8].

Analysis of Flow

A theoretical solution of the flow shown in Fig.1 can be obtained by dividing the circulation zone into two regions, MM' being the dividing section[9]. In the analysis the following assumptions are made:

(1) There are no external forces such as pressue gradient or wall shear acting on the flow.

(2) In the first region of flow, the velocity profiles in the mixing layer are assumed to be similar and are given by Schlichting's formula,

$$\frac{U_1 - u}{U_1 - U_2} = (1 - \eta^{1.5})^2$$

where $\eta = (y - y_2) / (y_1 - y_2)$.

(3)   The growth of the mixing layer width is assumed linear, the same as when a free air jet discharges into the atmosphere:

$$b = (y_1 - y_2) = Cx = 0.3x$$

The value of the coefficient C was found by experiments for the flow behind a two-dimensional bluff body.

(4)   At section MM' it is assumed that the average kinetic energies in the forward and reverse flow and also their cross-sections are equal.

The results of the analysis for the rates of spreading in the first region and for the variation of the reverse velocity are shown in Figs. 3 and 5. The analysis also gives

$$l_1 = 5.1\ B, \quad l_2 = 0.9\ B \ and\ m_{max} = -0.4$$

where B = 0.5 (B_2 - B_1) and m = U_2/U_1. Hence the total length of the circulation zone is

$$L_T = 6.0\ B$$

Experimental Equipment

The experiments were carried out in a 1ft wide glass flume having an outlet weir for adjusting the flow depth.

The flume was fitted with perspex guide walls to produce a stream 3 in. deep by 6 in. wide with a Froude number F_1 = 0.5, expanding abruptly on one side into a trapezoidal channel with a side slope of 1 vertical to 2 horizontal.

For measurements of velocity, a shielded dynamic probe, 0.375 in. external diameter, was used[4]. The probe was found to be insensitive to direction changes through a range of angles of $\pm$ 40 degrees. The pressure distribution at any section was assumed hydrostatic and was locally measured by a point gauge graduated in 0.002 in.

Generally, the distribution of velocity in an open channel is not uniform; the maximum value usually occurs between 0.05 and 0.25 of the depth h below the free surface. In the present experiments, velocity profiles were measured at 0.2h below the water surface.

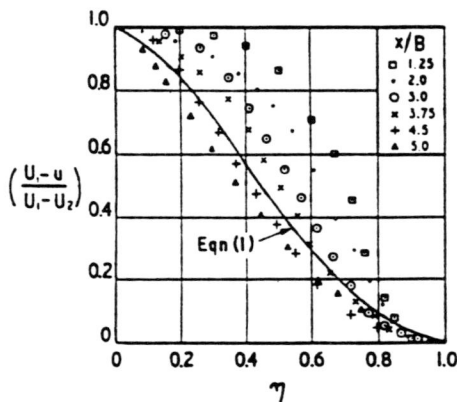

Fig. 2—Comparison of velocity profiles

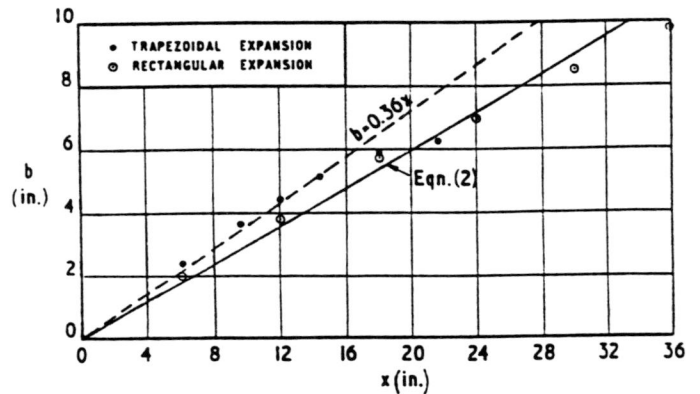

Fig. 4—Width of mixing layer

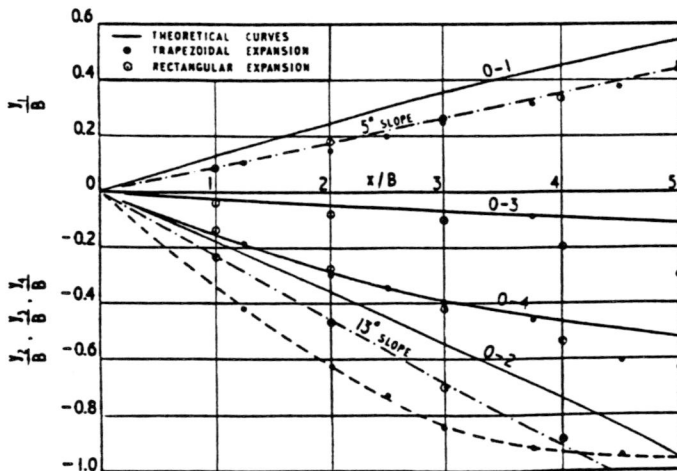

Fig. 3—Spreading of stream boundaries

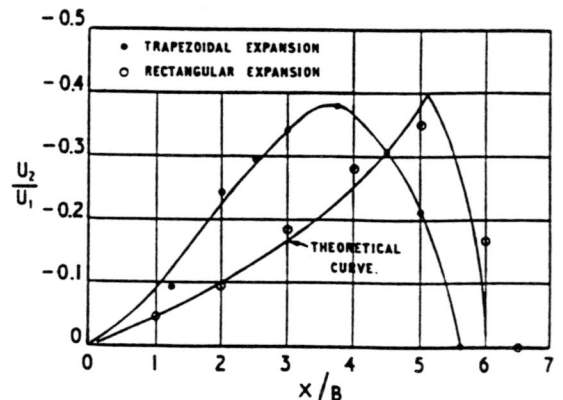

Fig. 5—Variation in velocity of reverse stream

Discussion of Results

The velocity profiles measured at various sections across the stream are plotted non-dimensionally in Fig.2, together with the three-halves power law of Schlichting. The profiles are seen to vary from section to section and are, therefore, dissimilar; the actual velocity falls more slowly on the stream edge but faster on the circulation side than given by Schlichting's curve. For the rectangular expansion[4], the velocity profiles were found approximately similar only for $x/B \geq 4$.

The experimental and theoretical results for the boundaries of the stream are shown in Fig.3. The chained lines show the corresponding results for the rectangular expansion. The inner boundary is seen to expand at the same angle of 5 degrees or 1 in 12 as for the rectangular case, which is less than the predicted rate for 01 but it is the same as that for a half jet analyzed by Tollmien using Prandtl's mixing-length theory[10]. The spreading of the outer boundary is not linear but follows a curve, the slope of which indicates a much greater rate of spreading than that for 02.

At section $x/B = 3.75$, the outer boundary joins to the wall layer on the side of the channel, thus extending the turbulent mixing layer across the full width of the circulation zone. The overall width of the mixing layer b is plotted against x in Fig.4; the results lie closely to a line with a slope of 0.36 which is appreciably greater than that given by Equation (2).

Fig.5 shows the variation in velocity of the backflowing stream plotted on the theoretical curve, together with the results for the rectangular expansion. The maximum reverse velocity agrees well with the theoretical value of $0.4 U_1$, but it occurs at $x/B = 3.75$, the section where the outer boundary reaches the wall JN. This section also defines the downstream end MM' of the first region of the circulation zone. The length of this region is 3.75 B compared with the predicted value of 5.1 B.

In the second region of the circulation zone, fluid must be recirculated from the stream to replace that entrained in the first region. Apparently, due to this recirculation and also due to the limited mixing width as the outer boundary reaches the wall JN, the mixing process is intensified, which results a rapid broadening of the stream until it occupies the whole width of the trapezoidal channel at the stagnation point, which defines the downstream end of the circulation zone. The length of the second region is 1.85 B, which is practically twice the theoretical value. The total length of the circulation zone is 5.6 B compared with the theoretical value of 6.0 B and that for the rectangular expansion of 6.5 B.

It should be noted that due to the side slope in the trapezoidal expansion the length of the circulation varies with depth, ranging from zero to the bed to a maximum at the water surface. This variation in length could, to some extent, account for the faster rate of spreading than in the rectangular expansion, and also for the departure of the results from theory.

Design of Transition Protection

When a rectangular stream is discharged abruptly into a trapezoidal earth channel the emerging flow velocity is normally higher than the maximum permissible value for the soil formation, and if the channel bed is unprotected scour will occur. Furthermore, the interface between the stream and the circulation fluid consists of a series of large scale eddies which, because of their strong vortex motion, erode the channel banks and destroy their stability. To safeguard against these two actions, a protection must be provided in the form of stone pitching or concrete blocks or both overlying an inverted filter as shown in Fig.6. The length of the protection is fixed by the length of the circulation zone, which for a slope of 1 : 2 is about eleven times the depth of flow.

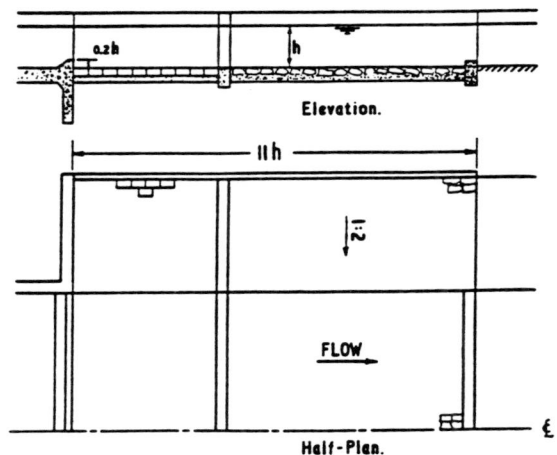

Fig. 6—Design of protection downstream of channel expansions

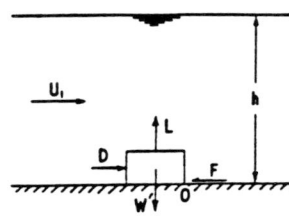

Fig. 7—Forces on a prism resting on bed of a stream

To calculate the size of the stone or block, consider a prism of density ρ_1, resting on bed of stream with mean velocity U_1, Fig.7. As the flow passes the block, the streamlines are deflected upward to form a wake downstream. Due to the change in flow pattern, the prism is subjected to a drag force D in the direction of flow and a lift force L perpendicular to it. These hydrodynamic forces are commonly expressed by the following equations:

$$D = C_D \frac{1}{2} \rho b d\, U_1^2$$

$$L = C_L \frac{1}{2} \rho b l\, U_1^2$$

where
b =	width of prism across the flow
l =	length of prism parallel to flow
d =	depth of prism
ρ =	density of water
C_D =	coefficient of drag
C_L =	coefficient of lift

The forces resisting the motion of the prism are the submerged weight W^1 and the sliding friction force F between the prism and the channel bed:

$$W' = b d l g\, (\rho_1 - \rho)$$

$$F = \mu\ (W' - L) = \mu\ b d l g\, (\rho_1 - \rho)$$

$$- \mu\ C_L \frac{1}{2} \rho b l\, U_1^2$$

μ being the coefficient of friction. The movement of the prism will depend upon the relative magnitudes and interaction of the above forces and may occur either by sliding or by overturning and rolling if initially it rests against a projection in the bed.

For equilibrium against movement by sliding, the drag force must be resisted by the sliding friction force,

$$C_D \frac{1}{2} \rho b d\, U_1^2 = \mu\ b d l g\, (\rho_1 - \rho)$$

$$- \mu\ C_L \frac{1}{2} \rho b l\, U_1^2$$

which reduces to

$$l = \frac{\left(\dfrac{C_D}{\mu} + C_L \dfrac{l}{d} \right)}{\left(\dfrac{\rho_1}{\rho} - 1 \right)} \frac{U_1^2}{2g}$$

For equilibrium against movement by overturning and rolling along the bed, the sum of the moments of lift and drag forces about a line through the edge 0 must equal the moment of the submerged weight about the same line. That is:

$$C_D \frac{1}{2} \rho b d\, U_1^2\ \frac{d}{2} + C_L \frac{1}{2} \rho b l\, U_1^2\ \frac{l}{2}$$

$$= b d l g\, (\rho_1 - \rho)\ \frac{l}{2}$$

which gives

$$l = \frac{\left(C_D \dfrac{d}{l} + C_L \dfrac{l}{d} \right)}{\left(\dfrac{\rho_1}{\rho} - 1 \right)} \frac{U_1^2}{2g}$$

Comparing Equations (9) and (11), it is clear that for a given velocity the required equilibrium length of block is greater for sliding than for overturning, consequently the design of the blocks should be based on Equation (9).

The values of C_D and C_L were found experimental[11] by measuring the lift and drag forces on plasticine prisms held about 1 mm above the bed of a glass flume and are plotted in Fig.8. The measurements were taken for various discharges and depths ranging from 4 to ^XxV. For depths h < 4d, the value of C_D was found to increase rapidly reaching 2.0 for h = 2.3d. Over the same range, the value of C_L was practically zero.

Friction tests carried out by sliding stones and concrete blocks on a smooth concrete surface with a cord and scale pan indicated an average value for μ of 0.3. Assuming this and square blocks with l/d = 3, Equation (9) becomes

$$l = \frac{(C_D/0.3 + 3\ C_L)}{(\rho_1/\rho - 1)} \frac{U_1^2}{2g}$$

Given the unit discharge and depth of flow, the velocity $U_1 = q_1/h$ and the Froude number $F_1 = U_1/(gh)^{1/2}$ are first calculated, and then the values of C_D and C_L are obtained from the curves in Fig.8. Substituting these values in Equation (12), the stable length of the stone or block can be determined. The other dimensions of the protection are obtained from Fig.6.

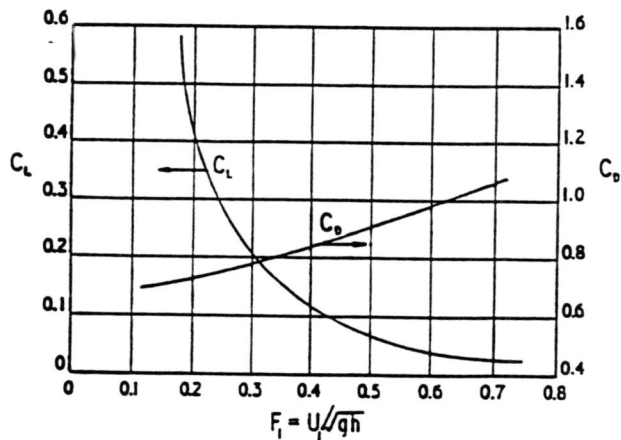

Fig. 8—Variations of lift and drag coefficients for $h \geqslant 4d$

References

1. Naib, S.K.A. "Flow patterns in a submerged liquid jet diffusing under gravity", Nature, Vole.210, p.694, May 14, 1966.

2. Naib, S.K.A. "Photographic method for measuring velocity profiles in a liquid jet", The Engineer, Vol.221, p.961, June 24, 1966.

3. Naib, S.K.A. "Measuring velocity in a liquid jet" Letter to the Editor, The Engineer, Vol.222, p.236, Aug.12 1966.

4. Naib, S.K.A. "Mixing of a subcritical stream in a rectangular channel expansion", Journal of the Institution of Water Engineers, Vol.20, p.199, May 1966.

5. Naib, S.K.A. "Unsteadiness of the circulation pattern in a confined liquid jet", Nature, Vol.212, p.753, November 12 1966.

6. Naib, S.K.A. "Abnormal groupings of large eddies in a submerged jet", La Houille Blanche, No.3, p282, 1967.

7. Naib, S.K.A. "Analysis of subcritical channel transitions", Water and Water Engineering, Vol.71, p.55, February 1967.

8. Naib, S.K.A. "Hydraulic design of energy dissipators", Water and Water Engineering, Vol. 70, p.191, May 1966.

9. Abramovich, G.K. "The theory of turbulent jets", M.I.T. Press, Boston, Mass., USA, 1963.

10. Tollmien, W. "Calculations of turbulent expansion processes", NACA Tech. Memo. No.1085, 1945.

11. Naib, S.K.A. "Equilibrium of talus blocks downstream of stilling basin", Water Power, Vol.19, p.406, October 1967.

PART V

JET DISPERSION
IN CHANNELS

Submerged Pipe Outlets
Jet Dispersion in Channels
Jet Dispersion in Confined Spaces

SUBMERGED PIPE OUTLETS

Summary

This experimental investigation concerns a high velocity jet issuing from a submerged pipe outlet and diffusing into a rectangular channel in order to explore the dissipation of velocity for various inclinations of the jet. Particular attention is paid to the vertical jet while varying the height of impingement, channel width, downstream depth and Froude number of the flow. The results are compared with data from previous experiments on air jets impinging on a flat plate and a generalised power law is derived to enable the deceleration of the jet to be predicted, a matter which obviously bears upon the design of stilling basins for outlets of pumping stations, spillways and tunnels and irrigation and drainage works.

Nomenclature

a =	Intersection of central axis of pipe outlets with channel bed	
C =	Proportionality coefficient	
d =	Pipe diameter	
F_o =	Froude number of flow at pipe outlet $U_o/\sqrt{gd}$	
H_t =	Maximum total head	
h_t =	Total head	
h_1 =	Height of Impingement	
h_2 =	Downstream depth	
l_w =	Lateral width of jet from centreline	
n =	Exponent of power law	
R_o =	Pipe exit Reynolds number = $U_o d/\nu$	
U_o =	Pipe exit velocity	
U_1 =	Initial maximum velocity of deflected jet	
U =	Maximum centreline velocity	
w =	Width of channel	
x =	Distance along centreline	
ϕ =	Angle of Impingement	
ν =	Kinematic viscosity of the fluid	

Introduction

Jet dispersion in an ambient fluid of infinite extent is a problem of fluid motion on which a considerable amount of experimental and theoretical investigation has been published [Ref 1,2]. The more complex motion of a round jet dishcarging from a pipe outlet directly into a rectangular channel, however, has received far less attention, and no records were found of fundamental work on the behaviour of such a jet or the effect of the depth and width of the channel on the dissipation of a jet entering it; a matter which obviously bears upon the design of stilling basins for outlets of pumping stations, spillways and tunnels and irrigation and drainage works. In practice the same problem arises, namely the dissipation of the energy of the jet, in such a manner as to provide smooth flow with a sufficiently low velocity of flow as will not cause scour in the downstream channel, or damage to the outlet structure.

Although design methods are available for some types of jet dissipation basins [Ref 3], these are based on empirical studies of specific cases without any general theory and analysis to enable design to be extended to other situations. It is the generalisation of information arising from a wide ranging but fundamental study, such as is now investigated that is considered to be neccesary. The research follows and is based on previous investigations. Experiments were initially carried out on a round air jet projected parallel to a wall [Ref 4] and at angles of 15°, 30° and 45° to a plane smooth surface [Ref 5]. For the 30° and 45° deflected jets the decay of maximum velocity and head loss were found to be much faster than for unconfined submerged jets.

Experiments were then conducted on a jet issuing from a submerged horizontal pipe into an open channel. It was observed that the jet disperses with intense turbulence and asymmetry and is extremely unstable. Attempts to stabilize the jet without altering its character, were unsuccessful. However, a reasonably steady and symmetrical flow was obtained when the jet was directed at about 45° and more the channel and ran up the side walls very smoothly, and much of its excess energy dissipated at a short distance downstream [Ref 6].

The main aims of the current research were to carry out investigations into oblique jet dispersion in channels (Fig 1), in order to determine an efficient form of stilling basin for high velocity discharge from a pipe outlet and the minimum size of the basin for any given flow conditions. In the experiments the inclination of the jet was varied from 45° to 90° to the bed of the channel, while differing ratios of limiting downstream depth to jet diameter were studies for various Froude numbers. The effects of the jet height above the bed, the width of channel and jet distance from the upsteam face of the channel were also investigated. This technical note reviews some of the laboratory work to date and presents data to enable designers to predict the deceleration of the jet. Detailed measurements of the flow characteristics and the development of generalised procedures for the design of stilling basins are still in progress.

Apparatus and Procedure

The work was carried out in special water jet equipment consisting of a pipe discharging with varying inclinations into a glass flume 1800 mm long and 300 mm wide. Perspex walls were inserted into the channel to vary its width. The flow was supplied by a centrifugal pump and measured by a digital meter.

Measurements of the maximum central velocity U were initially obtained by the use of a Pitot-static tube. However, it soon became evident that the boundary layer was actually thinner than the diameter of the Pitot-static tube itself. The determination of the velocities therefore, had to proceed by a different measurement technique. The static head was still obtained by the use of the Pitot-static tube but the total head was recorded by a total head tube with an inside diameter of 1 mm. The range of experiments performed are summarised in Table 1. The symbols appearing in this table are defined in Fig.1.

Subsequently, the Laser Doppler Anemometry (LDA) technique was adapted for measurements of mean velocities and turbulence characteristics without any interference in the flow. The use of a Bragg Cell in the system made instantaneous measurements of reverse flows in the recirculation region possible. Information from these units were fed into a PET computer where the mean and RMS value of the velocity was measured. Measurements confirm the results obtained by the total head tube.

The discussion of the results obtained is divided into

sections dealing with each of the oblique and normal jet impingements tested and it is convenient to comment on the various phenomena with reference to the particular flows for the normal jet impingement, but some of the remarks made, also apply to the oblique case.

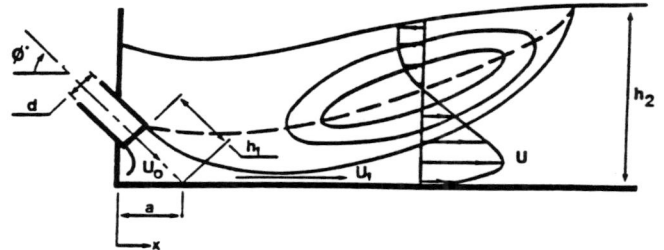

Fig 1   Flow downstream of submerged pipe outlet

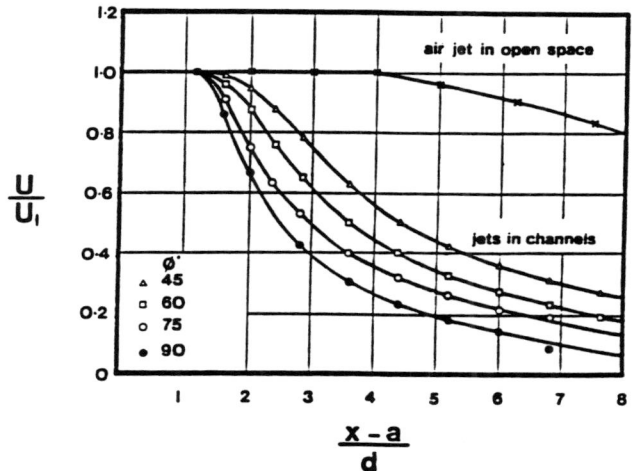

Fig 2   Variation of velocity with impingement angle

Analytical Background

The design of stilling basins, like the design of many hydraulic structures involving jets whose expansion is limited by complex boundary conditions, will probably for a long time to come be based on the results of experimental tests. The details of the flow are so complicated that no known theory could describe them fully at the present time. However, the main features of certain jets have been analyzed, and the results while not directly applicable are of great help in explaining the motion of the jet from a pipe outlet.

Using Prandtl's mixing length theory, Tollmien solved the fundamental equations of turbulent motion of a round jet originating from a point source and diffusing into ambient surrounding fluid and showed that the central velocity of the jet U is inversely proportional to the distance x, i.e.

$(U/U_1) = C(d/x)$, where U_1 is the initial velocity of the jet, C is a coefficient for the jet and d is the orifice diameter. The value of C has been found experimentally to be variable depending on the effect of the boundary layer in the nozzle. Based on theoretical and experimental results a mean value of 6.5 has been recommended. Kuethe extended Tollmien's work to the case of a jet with a finite source and established the length of potential core is about 5 diameters downstream of the orifice. The result agreed with experiments on free air jets over a wide range of Reynolds numbers, but unfortunately no comparable results were found on submerged water jets. For an extensive discussion of these works the reader is referred to References 1 and 2.

The flow of a round jet impinging vertically on a plate and spreading radially over it has been termed as the wall jet and analyzed by Glauert [Ref 7]. An experimental investigation of the radial wall jet has been carried out by Bakke [Ref 8]. Measurements of velocity profiles were made at distances 10 to 20 times the jet diameter. At these distances, the profiles were similar and the rate of decay of the jet followed the power law of $U (x)^{-1.11}$ which was in fair agreement with Glauert's prediction. Studies of the characteristics of the velocity and pressure fields in the jet have also been conducted by a number of other investigators [Refs 9, 10, 11, 12, 13, 14].

Oblique Jet Impingement in Channels

Impingement Angle

The results showing the variation of the maximum velocity U along the middle of the channel are plotted non-dimensionally against distance in Fig.2. The maximum velocity U_1 is the initial velocity of the deflected jet, which due to lateral spreading into a thin stream on impact, has a slightly greater value than the pipe exit veloicty U_o. The results of experiments of a round air jet discharging into the open atmosphere are also given. It is remarkable that at a short distance of 5 diameters downstream of the inclined pipe outlet the velocity decreases more than 15 times faster in the channel than in the submerged air jet. The maximum velocity in the channel is reduced to less than one-quarter in a distance of 8 diameters, compared to four-fifths for the jet in the open. Fig.2 also shows that with increasing jet inclination the rate of decay is increased reaching a maximum when the jet is discharging vertically into the channel. This is due to greater lateral spreading of the vertical jet on impact producing a large area which then becomes available for mixing and hence head dissipation.

The increase in lateral spreading of the jet on impact with increasing impingement angles were clearly visible under free flow conditions. These widths l_w were measured along the bed and sides of the channel for a Froude

number of 8 and are given in Table 2.

Power Law of Decay

To establish the law of decay in the deflected region of the jet a logarithmic plot of the results is shown in Fig.3. The graph shows three regions of flow. Region 1 of potential core extending to 1.25d followed by Region 2 of short transition up to 2.5d in which the velocity is gradually reducing and a final Region 3 where the flow appears to be fully developed and the results are governed by the following power law:

$$\frac{U}{U_1} = C \left(\frac{d}{x-a} \right)^n$$

where C is the proportionality coefficient and n is the exponent. The values of C and n are given in Table 3. For purposes of comparison data from previous investigations on similar air jets impinging on a flat plate are also included [Refs, 5, 8, 13, 14]. These values are plotted in Figs.4 and 5 to clearly show the variation of C and n with increasing ϕ. It is to be noted despite the variation in jet height, for the air jets, there is fair agreement in the values of C and n up to $\phi = 45°$.

For greater values of ϕ there is a marked increase in the value of n for the water jet in the channel. The range being 1.02 to 1.45. Observations of the water jet under free flow conditions show that this is due to part of the flow being deflected inward from the side and back walls, to the centre of the channel, cuasing a greater degree of mixing in the expanding jet and hence faster head dissipation.

The instabilitiy of the jet at angles of impingement less than 45° prevented any information being obtained in this range, however it appears from Fig.4 and Fig.5, if the data for the water jet were extrapolated for the rate of decay, for $\phi < 45°$, would be governed by the same relationship as for the air jet impinging on a flat plate, i.e. the side walls would have very little or no effect. This is understandable since at angles of impingement less than 45° the lateral spreading of the jet is not great enough to cause any significant deflection off the side walls of the channel.

An insight into the rate of dissipation of the total head based on the maximum centre line velocity, for varying angles of impingement, is given in Fig.6. It is seen that downstream of the potential core the rate of dissipation is very rapid up to a distance of 3.5d. Further downstream the rate decreases and finally 85% to 90% of the total head is dissipated with a distance of 8 diameters.

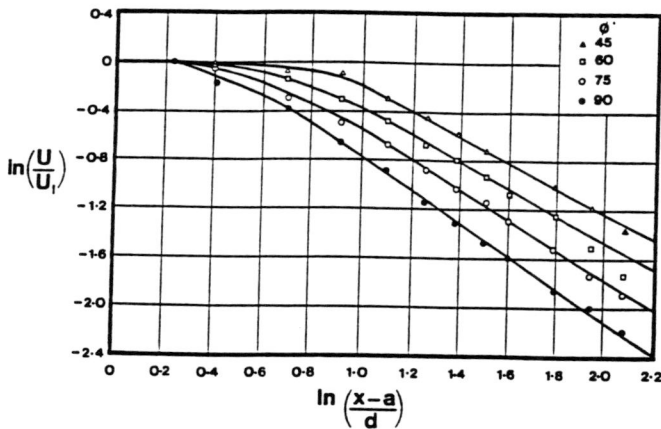

Fig 3   Establishment of power law

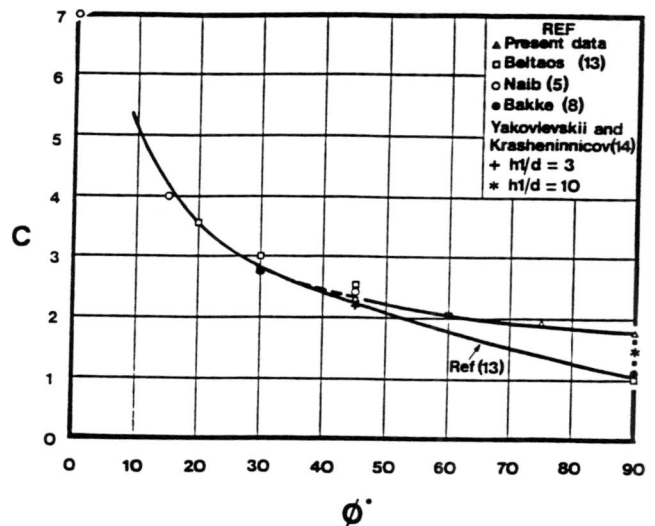

Fig 5   Variation of coefficient C with impingement angle

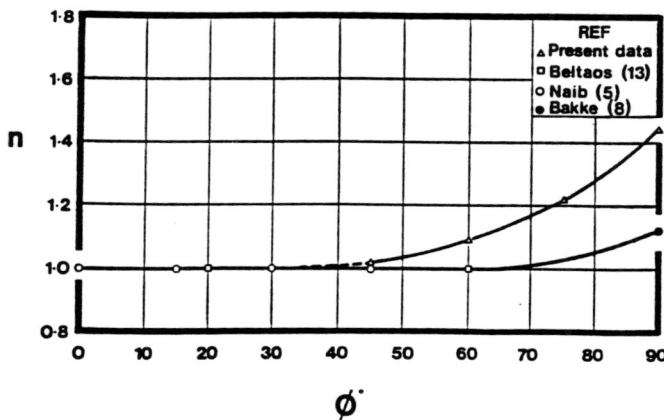

Fig 4   Variation of power n with impingement angle

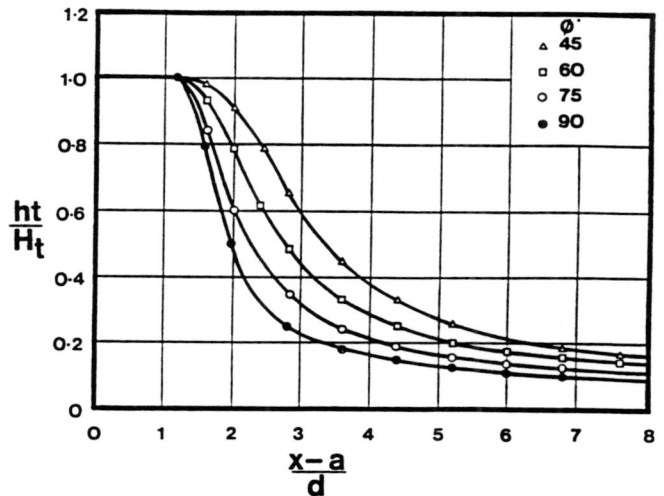

Fig 6   Dissipation of total head with distance

Vertical Jet Impingement in Channels

Jet Height

The results for maximum velocities for various jet heights and downstream water levels (Fig 7), seem to justify the hypothesis that increasing the jet height has little effect on the deceleration of the jet. This conclusion was also drawn by Beltaos [Ref 13] for an air jet impinging on a flat plate at heights ranging from 15d to 47d, and by Yakovlevski and Krashninnicov [Ref 14] for jet heights of 3.5d to 10d.

It was observed that on impact the jet was transformed into a thin stream, the thickness of which remained practically cosntant irrespective of its height as long as the flow was submerged. The measured velocities in the deflected stream were reduced by about 10% for an increase in height from 1d to 6d. This reduction is equivalent to the precentage loss in velocity for an air jet in open space over the same distance, Fig.2. The jet region about the point of impact is the potential core which is normally 5d downstream of the orifice.

Froude Number

Although the range of the test Froude numbers was limited between 4 and 9, it seems fairly certain that there is no tendency for the flow to change with the Froude number. The corresponding Reynolds numbers were in the range 10^4 to 10^5.

Downstream Depth

The experiments on raising the downstream water level from 3d to 8d at a constant Froude number and channel width 12d show that the jet is not greatly affected by the downstream depth: the trend is that with increasing depth the rate of decay is decreased, Fig.8. This may be explained that at smaller depths the water surface rises in the downstream direction which indicates the conversion of the kinetic head of the jet into potential head resulting

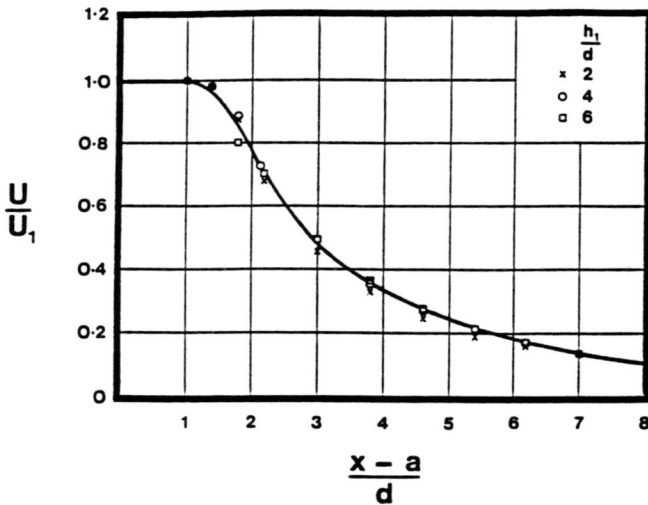

Fig 7   Effect of jet height

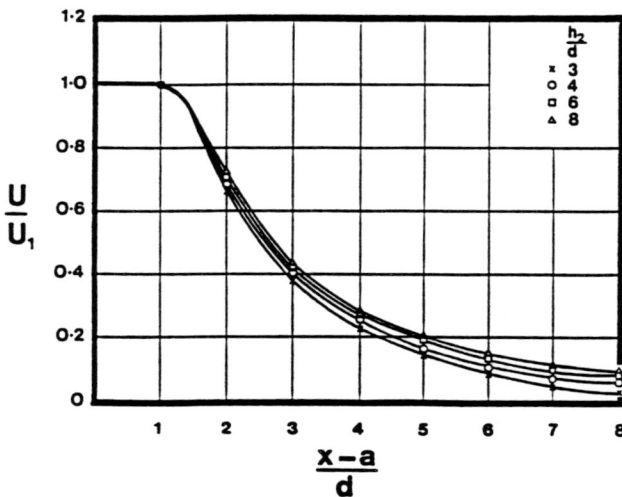

Fig 8   Effect of raising downstream water level

is increased, the amount dependent primarily on the width and depth of flow. For a width of 6d and depth variation of 4d to 6d the percentage rise ranges from 18% to 5%.

Position of Pipe Outlet

When the pipe is positioned flat against the back wall of the channel the upstream flow is eliminated and the flow towards the side walls is increased. This change in flow pattern at impact is characterised by slightly faster decay rate. As the pipe outlet is moved away from the back wall of the channel i.e. the dimension 'a' is smallest and the water surface becomes flatter with increasing 'a'. This appears to explain the slight variation in the degradation of the jet.

Conclusions

A study of the characteristics of the maximum centre line velocity of a round turbulent jet discharging obliquely into a rectangular channel has shown the following results.

(i)   Despite the complexity of the flow pattern in the deflected jet, three regions are apparent: region of potential core, region of transition flow and region of fully developed flow.

(ii)   The decay rate of the jet in the fully developed region is inversely proportional to a power n of the distance. The value of n is dependent on the angle of impingement and varies from 1.02 for $\phi = 45°$ to 1.45 for $\phi = 90°$ under the present experimental conditions. The maximum rate is achieved when the jet impinges vertically onto the bed of the channel.

(iii) The maximum velocity is reduced to less than one quarter in a distance of 8 diameters, which is equivalent to a dissipation of more than 85 per cent of the total head of the jet. Over the same distance, the head decreases more than 15 times faster as in the open air jet.

(iv)   The dissipation of head is increased by decreasing the width of the channel, keeping the downstream depth to a minimum of 4 diameters and positioning the pipe outlet approximately one diameter from the upstream end of the channel.

(v)   Raising the height of the jet above the channel bed has little effect on the decay rate and the tendency of increased height is towards decreasing the impact velocity at the core of the jet.

(vi)   Over the range of 4 to 9, surprisingly, the Froude

In faster decay. For high downstream levels there is no appreciable rise of the water surface. Observation of the flow shows a complex and multiple circulation on top of the deflected jet. The position of the stagnation point appears to be constant, between 4d to 5d, for a wide range of flow conditions, apparently due to the deflection of the jet from the side walls towards the centre of the channel.

Width of Channel

Variations in the width of the channel between 4d and 12d surprisingly showed only a small change in the decay rate of the maximum velocity. As expected the rate of decay is greatest for the channel of narrow width. It should be noted however, that as the channel width is decreased the depth of water immediately downstream of the pipe outlet number has practically no effect on the deceleration of the jet.

Applying these results to the practical design of a stilling basin, the indications are that maximum decay is achieved when the jet impinges vertically into a channel and that 90 per cent of its head it lost in a distance of a 8 diameters downstream. This allows for a shorter length of basin than that required for a horizontal outlet and can produce a saving in its length of about 50%.

References

1. Abramovich, G N: "The Theory of Turbulent Jets". The M.I.T. Press, 1963.
2. Rouse, H: "Advanced Mechanics of Fluids". John Wiley and Sons, 1965.
3. Bradley, J N & Peterka, M: "Hydraulic Design of Stilling Basins: Small Basins for Pipe or Open Outlets". Proc. Paper 1406, ASCE, Journal of Hydraulic Division, Vol 83, No H75, Oct. 1957.
4. Naib, S K A: "Spreading and Developmet of the Parallel Wall Jet". Aircraft Eng., pp 30-33, Decmeber 1969.
5. Naib, S K A: The Deflexion of a Submerged Round Jet to Increase Lateral Spreading". La Houille Blanche, Vol 29, No 6, pp 455-461, October 1974.
6. Naib, S K A & Sanders, J E: "Diffusion of Bluff Wall Jets in Finite Depth Tailwater". Discussion Paper 18354-HY, ASCE, Journal of the Hydraulics Division, February 1985.
7. Glauert, M B: "The Wall Jet". Journal of Fluid Mechanics, Vol 1, pp 625, 1956.
8. Bakke, P: "An Experimental Investigation of a Wall Jet". Journal of Fluid Mechanics, Vol 2, pp 212 (1957).
9. Poreh, M & Cermak, J E: "Flow Characteristics of a Circular Submerged Jet Impinging Normally on a Smooth Boundary". Proc. Sixth Midwestern Conference on Fluid Mechanics, Univ. of Texas, pp 198-212, 1959.
10. Bradbury, L J S: "The Impact of an Axisymmetric Jet on to a Normal Ground". Aeronautical Quarterly, Vol 23, pp 141-147, May 1972.
11. Beltaos, S & Rajaratnam, M: "Impinging Circular Turbulent Jets". Journal of the Hydraulics Division, ASCE, Vol 100, No HY10, pp 1313-1328, October 1974.
12. Davanipour, T & Sami, S: "Short Jet Impingement". Journal of the Hydraulics Division, Vol 103, NO NY5, pp 557-567, May 1977.
13. Beltaos, S: "Oblique Impingement of Circular Turbulent Jets". Journal of Hydraulic Research, Int. Ass. for Hydraulic Research, Vol 14, No 1, pp 17-36, 1976.
14. Yakovlevskii, O V & Krasheninnicov, S Y: "Spreading of a Turbulent Jet Impinging on a Flat Surface". Fluid Dynamics, Vol 1, No 4, pp 136-139, July-August 1966.

Table 1   Range of Experiments

Angle of Impingement ($\phi°$)	Jet Height (h_1/d)	Channel Width (w/d)	Downstream Depth (h_2/d)	Jet Distance Downstream (a/d)	Froude Number (F_0)	Reynolds Number ($R_0 \times 10^4$)
45,60,75,90	1	12	4	2	8	8.7
90	0.25,0.5,1,2,3	12	4	1	8	8.7
90	1,3,5	12	6	1	8	8.7
90	2,4,6	12	8	1	8	8.7
90	1	12	4	1	4,6,8,9	4.3,6.5,8.7,9.8
90	1	12	3,4,6,8	1	8	8.7
90	1	4,6,8,10,12	4	1	8	8.7
90	1	12	4	0.5,1,1.5,2,3	8	8.7

Table 2 Lateral spreading of Jet on Impact

Angle $\phi°$	15	30	45	90
Width ℓ_w	4d	6d	7.5d	10d

Table 3   Variation of C and n with impingement angle

Reference	Experiment	Jet Height (h/d)	Impingement Angle (ϕ)	C		n
Naib & Sanders (Present data)	Water jet in channel	1 to 6	45	2.32		1.02
			60	2.05		1.09
			75	1.97		1.22
			90	1.79		1.45
Naib (5)	Air jet on flat plate	<1	0	7		1
			15	4		1
			30	2.8		1
			45	2.4		1
Bakke (8)	Radial Wall Air Jet	0.5	90	1.12		1.12
Beltaos (13)	Air jet on flat plate	15 to 47	20	3.8		1
			30	3.0		1
			45	2.5		1
			60	2		1
Yakovlevskii & Krasheninnicov (14)	Air jet on flat plate	3.5 & 10		$\frac{h_2}{d}=3.5$	$\frac{h_2}{d}=10$	
			30		2.8	1
			45	2.3		1
			60	2.0	2.0	1
			90	1.2	1.5	1

JET DISPERSION IN CHANNELS

Introduction

Jet dispersion in an ambient fluid of infinite extent is a problem of fluid motion on which a considerable amount of experimental and theoretical investigation has been published[1,2]. The more complex motion of a round jet discharging directly into a rectangular channel, however, has received far less attention, and no records were found of fundamental work on the behaviour of such a jet, a matter which obviously bears on the design of stilling basins for outlets of spillway tunnels, and irrigation and drainage works. In practice the same problem arises, namely dissipation of the energy of the jet, in such a manner as to provide smooth flow with a sufficiently low velocity as will not cause scour in the downstream channel or damage to the outlet structure.

The US Bureau of Reclamation carried out tests and developed a generalised design for stilling basins of pipe outlets[3]. It is of the impact type with the jet impinging on a vertical wall and for incoming velocities less than 15 m/s. This design is suitable for irrigation outlets where no tailwater is required. For higher velocities there is violent turbulence in the flow and cavitation damage to the baffle wall may be considerable. Francis and Gunawardana[4] carried out measurements on a model of a drowned tunnel outlet in which a hump was used to improve the hydraulic performace. Their brief test at low Froude number of the flow was mainly directed at obtaining uniform flow distribution and determining energy losses. They concluded that detailed investigations were necessary of the many variables involved.

This paper describes the hydraulic research work carried out on a round turbulent jet discharging from a pipe outlet into a channel to study the degradation and spreading of the jet while varying the downstream depth of flow, width of the channel, height of jet and impingement angle. Based on these results and model studies, a generalised procedure has been evolved for the design of stilling basins for high velocity pipe discharge.

Apparatus

The work was carried out in special water jet equipment consisting of a 25 mm diameter pipe discharging with varying inclinations into a glass flume 1800 mm long and 300 mm wide. Perspex walls were inserted into the channel to vary its width. The flow was supplied by a centrifugal pump and measured by a digital meter. The depth of flow in the channel was controlled by means of an adjustable tailgate situated in the outlet tank.

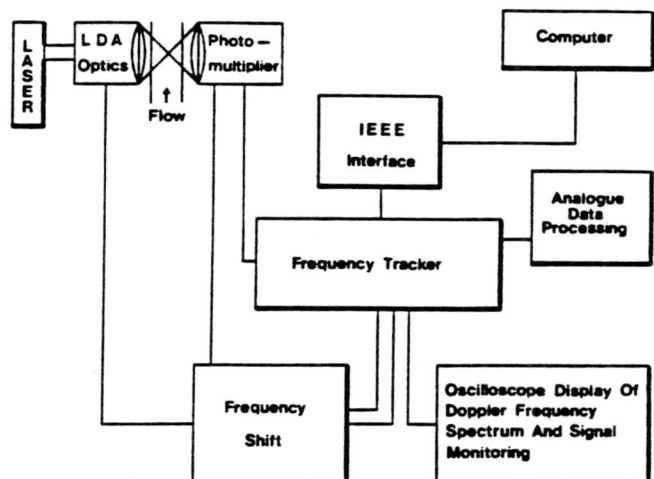

Figure 1

69

The Laser Dopler Anemometry (LDA) technique was adapted for measurements of mean velocities and turbulence characteristics, without any interference in the flow and utilized for the investigation of the complex three dimensional circulating flow which was experienced in the channel. Figure 1 is a schematic diagram showing the main components of the LDA system, as well as the data acquisition and analysis systems.

Oblique Jet Impingement

A broad programme of research was at first drawn for the investigation of a round jet issuing at various angles into a channel, but following the completion of certain preliminary experiments, it was considered that it would be more profitable to concentrate mainly on two cases of impingement, namely the 45° oblique jet and the 90° vertical jet.

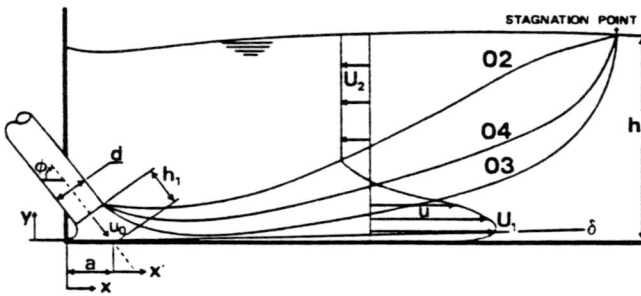

Figure 3

Jet Boundaries

Figure 3 shows the boundaries of the stream for a downstream depth of four jet diameters. The rate of growth of the boundary layer δ near the bed of the channel is linear with an inclination of 1°. The angles of spreading of the boundaries 02 and 04 are approximately 7° and 5.5° respectively.

The rate of growth of the normal half width $b(U_{(b)} = U_{1/2})$ is linear and follows the relationship $b = mx^1$. It was found that $m = 0.067$ which corresponds to an angle of growth of 3.8°. This was constant for a range of downstream depths between 4d and 8d. For an air jet impinging on a smooth flat surface at an angle of 45° Naib[5] suggests a value of $m = 0.086$ (4.9°) while the experiments of Yakovlevskii and Krasheninnicov[6] show b to grow linearly with a slope of 0.075 (4.3°).

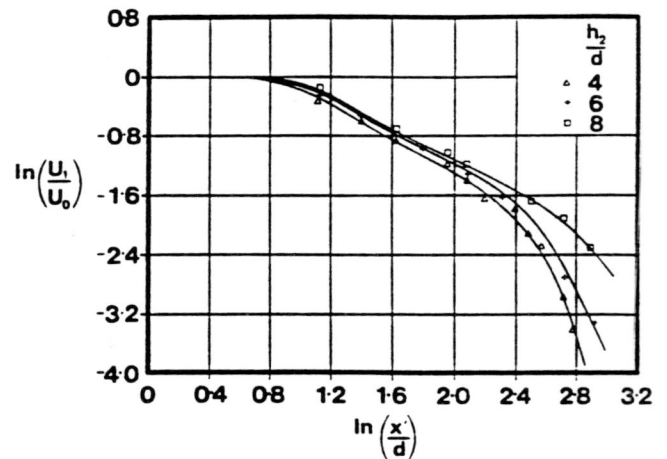

Figure 2

Flow Pattern

A schematic section of the motion of the oblique jet along the centreline of the channel is shown in Figure 2.

Immediately after issue the jet fans out on the floor of the channel. Between the fast stream and the surrounding fluid a turbulent mixing layer is formed having an outer boundary 02. The line 03 defines the mean flow line with the stream function $\psi = 0$, bounding the fluid which is sucked outward from the stream. The mixing process between the jet and the surrounding fluid causes a circulation zone to be established on top of the stream, in a similar manner to the flow behind a two-dimensional bluff body. The purpose of the investigation was to establish several parameters, including the rate of spreading, the decay of maximum velocity, the distribution and shape of the velocity profiles under different degrees of submergence. The results obtained are summarized below.

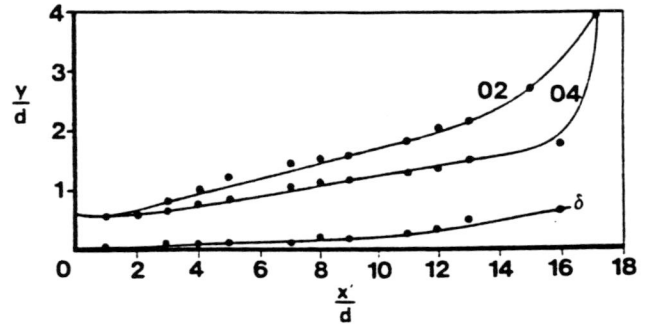

Figure 4

Maximum Velocity Decay

The rate of decay of the maximum central velocity U_1 is plotted logarithmically in Figure 4. The flow is fully developed after a distance of 3d; thereafter its decay rate is of the form shown below:

$$\frac{U_1}{U_o} = C\left(\frac{d}{x'}\right)^n$$

where C is a proportionality constant and n is an exponent.

For a downstream depth of 4d the values of these constants compare well with earlier work carried out by the authors[7]. In this region of flow the rate of decay is also similar to those found by Naib[5], Beltaos[8] and Yakovlevskii and Krasheninnicov[6] as shown in Table 1.

AUTHORS	C			n
	h_2/d 4 6 8			h_2/d=4,6,8
Naib/Sanders (present data)	2.4 2.6 2.7			1.06
Naib[5]	2.4			1
Yakovlevskii and Krasheninncov[6]	2.3			1
Naib/Sanders[7]	2.32			1.02
Beltaos[8]	2.5			1

TABLE 1

Following the entrainment region a transition takes place into the recirculation region where the jet decays much faster and resembles uniform channel flow shortly downstream of the end of the circulation zone. Seventy percent of the original velocity is lost within a distance of 8d and an additional 20% over the following 12d.

Decreasing the downstream depth changes the overall rate of decay. The maximum velocity remains inversely proportional to the distance, with n = 1.06 but the value of the proportionality constant C decreases thus indicating a shift in the virtual origin. This is the result of the presence of greater pressure gradients when the downstream water level is low, indicating a conversion of some of the kinetic head to static head.

In Figure 5 the non-dimensional plot of the velocity profiles for distances between 3d to 7d lie nearly on one curve and are therefore similar. The end of similarity of velocity profiles indicates the termination of the entrainment region and a change in the rate of decay.

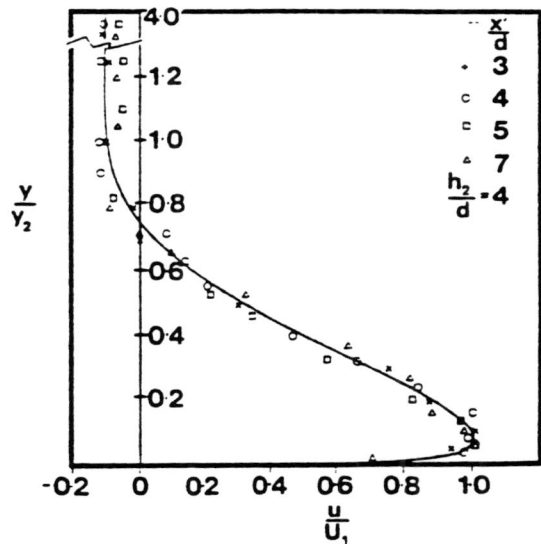

Figure 5

Vertical Impingement

Flow Pattern

The flow pattern illustrated in the definition sketch Figure 6 for vertical impingement clearly indicates a complex multiple-circulation region above the deflected jet.

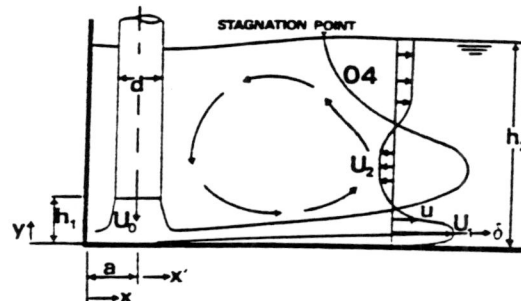

Figure 6

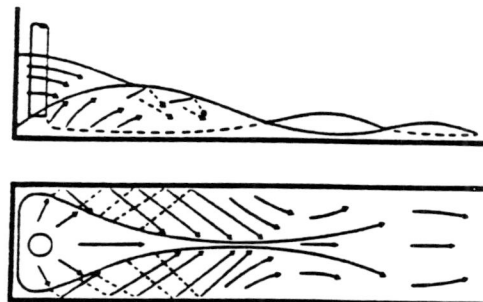

Figure 7

71

It is apparent that the upper half of flow has been forced to move downstream instead of re-circulating as in the case of the 45° impingement. Observations of the free flow conditions (figure 7) show that on impact the deflected jet spreads along the bed and up the side walls of the channel.

The heights of the stream on the side walls is dependent on the width of channel and the result of this is the flow being deflected and falling back into the centre of the channel. When the jet is submerged it is the fluid deflected from the side walls which causes the upper part of the circulation region to be forced into flowing downstream.

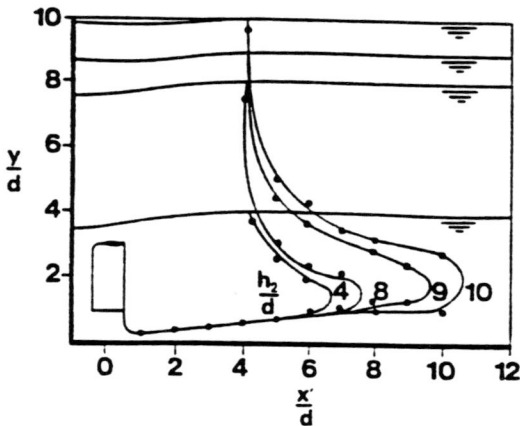

Figure 8

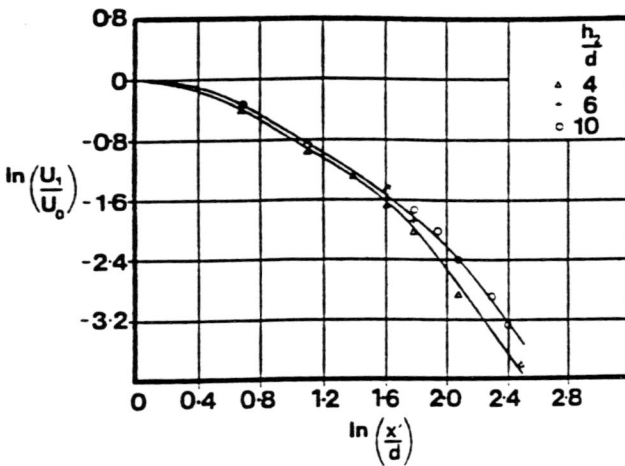

Figure 9

Jet Boundaries

The stream boundary 04 for different downstream depths is plotted in Fig.8. The mean angle of spreading of this boundary is 9°.

It is interesting to note that the stagnation point remains constant at about 4d for the range of depths tested. This is again the result of the flow deflected from the side walls. The geometry of the deflected jet is such that it remains the same for all depths of submergence, therefore the flow from the side walls always falls back into the centre of the channel at the same position, hence the position of the stagnation point remains practically constant. The length of the submerged part of the circulation region is prevented from expanding normally to the surface by the flow from the side walls.

The rate of growth of the normal half width b for a deep submergence of 10d, was found to be $b = 0.08x^1$ (4.6°). This is surprisingly in close agreement with the result of an air jet impinging vertically on a flat plate[9,10,11].

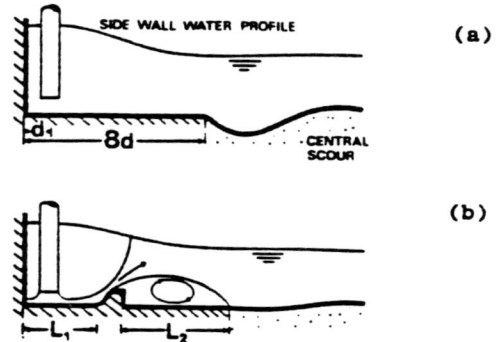

Figure 10

Maximum Velocity Decay

Figure 9 gives examples of the distribution of maximum velocity along the centreline of the channel for various downstream depths. For the entrainment region, $2d < x^1 < 5d$, the variations are linear and may be represented by the following equations:

$$\frac{h_2}{d} = 4 \; ; \; \frac{U_1}{U_o} = 1.6 \left(\frac{d}{x^1}\right)^{1.3}$$

$$\frac{h_2}{d} = 6, 8, 10 \; ; \; \frac{U_1}{U_o} = 1.7 \left(\frac{d}{x^1}\right)^{1.3}$$

72

These results show some deviation from earlier work by the authors[7] where $U_1/U_o = 1.79 \, (d/x^1)^{1.45}$ for $2 < (x^1/d) < 8$. This is thought to be due to the fact that in the earlier work carried out the decay was assumed to be at a constant rate between 2d and 8d. Whereas the current and more accurate measurements, which were made possible due to the utilisation of the LDA equipment, show that after 5d the rate of decay changes. This may indicate the end of an entrainment region, and a trasition into the recirculation region. The different rates of decay for varying depths may be due to the presence of a pressure gradient as evidenced by the slope of the water surface in the direction of flow. For depths greater than 6d the decay rate remains unchanged as the water surface becomes flatter at these depths and hence the change in pressure gradient is too small to have an effect on the flow.

Approximately 90 percent of the pipe outlet velocity is lost over a distance 8d compared with only 70 percent for the 45° impingement.

Jet Position

The results of tests for various jet heights seem to indicate that increasing the height of impingement has little or no effect on the deceleration of the jet[12]. However, increasing the height of the jet causes the water level immediately upstream of the pipe outlet to rise and therefore for practical considerations it should be kept small. A height of one jet diameter above the channel bed is recommended. The positioning of the pipe from the back wall of the channel is another important factor and again from maximum velocity and minimum scour considerations a distance of one jet diameter from the back wall was found to give the least scour downstream.

Froude Number

Experiments were also carried out for various Froude numbers and although the range was limited between 4 and 8, it seems fairly certain that there is no tendency for the rate of decay of the jet to change with Froude number. The corresponding Reynolds numbers were in the range of 10^4 to 10^5.

Stilling Basin Design

Applying the results of this experimental investigation to the practical desing of a stilling basin, the indications are that maximum decay is achieved when the jet impinges vertically into a channel and that 90 percent of its head is lost in a distance 8d downstream. These results prompted an initial design which is illustrated in Figure 10a. Preliminary scour tests showed this to be inadequate; so the length of the apron was extended and some improvement was noted.

A further series of experiments has then made in which a cill was placed on the apron, Figure 10b. Here, the jet was deflected away from the bed and a recirculation zone was established, resulting in a considerable reduction of the scour downstream. By varying the distance of the cill (L_1) from the back wall of the channel and studying the resulting scour profiles its optimum position was located. The length of apron required after the cill (L_2) was determined by analyzing the observations made of the recirculation region created behind the cill. This arrangement resulted in the elimination of practically all scour downstream of the stilling basin apron. The effect of the cill was to create a high rise in water level around the pipe outlet at low depths of submergence. This was simply overcome by depressing the length of apron before the cill giving the jet a greater volume of water above it. The pool suppressed the rise in water to an acceptable degree.

The tests were carried out under three different downstream depths and at four different widths. The maximum velocities were around 4 m/s with a corresponding Froude number $F_o = 8$. The values of L_1 and L_2 were found to be 4d and 10d respectively with a cill deflection angle of 45°.

The depth of pool required (p) for various channel widths (w) was determined by putting a constraint on the permissible rise in water level around the pipe otulet and the following relationship was obtained:

$$p = 8d - \left(\frac{w}{2}\right)$$

Based on test measurements, the height of the stilling pool water level, h_p, for downstream depths of 2d to 6d, may be estimated from the following equation:

$$h_p = 2d + 0.7 \, h_2$$

For greater submergence, h_p and h_2 are practically the same.

Figure 11 shows the proposed design for the stilling basin which has been developed from the series of model and scour tests and the investigation of the maximum velocity degradation described in this paper. Work is now underway for further refinement of the design procedure.

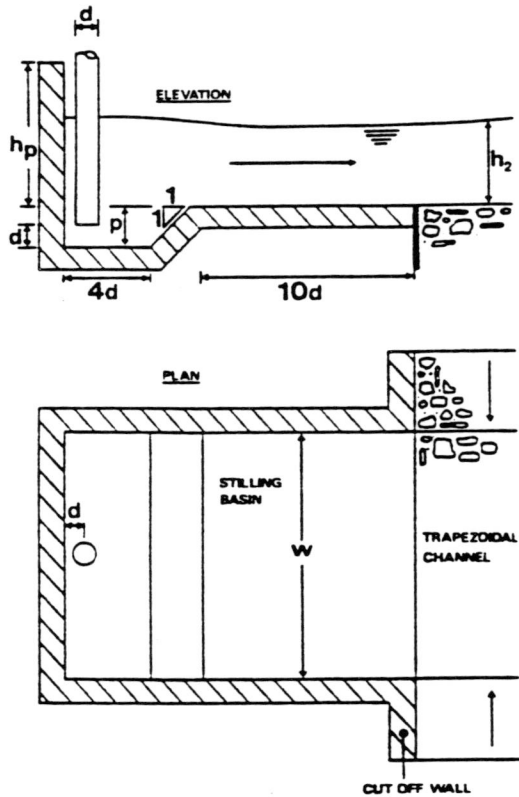

Figure 11

References

1. Abramovich, G N. "The Theory of Turbulent Jets". The MIT Press, 1963.

2. Rouse, H. "Advanced Mechanics of Fluids". John Wiley and Sons, 1965.

3. Bradley, J N & Peterka, M. "Hydraulic Design of Stilling Basins" Proc. Paper 1406, ASCE, Journal of Hydraulics Division, Vol 84, No H75, Oct 1957.

4. Francis, J R & Gunawardana, O A. "Hydraulic Design of Pipe Outlets into Open Channels". JICE, December 1969.

5. Naib, S K A. "The Deflexion of a Submerged Round Jet to Increase Lateral Spreading". La Houille Blanche, Vol 29, No 6, pp 455-461, October 1974.

6. Yakovlevskii, O V & Krasheninnicov, S Y. "Spreading of a Turbulent Jet Impinging on a Flat Surface". Fluid Dynamics, Vol 1, No 4, pp 136-139, July-August 1966.

7. Naib, S K A & Sanders, J E. "Jet Dispersion Downstream of Pipe Outlets". BHRA International Conference on the Hydraulics of Pumping Stations, Proceedings, Paper 10, Manchester, England. 17-19, September 1985.

8. Beltaos, S. Oblique Impingement of Circular Turbulent Jets". Journal of Hydraulic Research, Int. Ass. for Hydraulic Research, Vol 14, No 1, pp 17-36, 1976.

9. Bradshaw, P & Love, E M. "The Normal Impingement of a Circular Air Jet on a Flat Surface", ARC, R and M, No 3205, 1959.

10. Poreh, M., Tsuei, Y G ^ Cermak, J E. "Investigation of a Turbulent Radial Wall Jet". J. of Applied Mechanics, June 1967, pp 457-463.

11. Hrycak, P., Lee, D T., Gauntner, J W & Livingood, J N B. "Experimental Flow Characteristics of a Single Turbulent Jet Impinging on a Flat Plate". US NASA, TN-D-5690, 1970.

12. Rajaratnam, N. "Developments in Water Science; Turbulent Jets". Elsevier 1976, pp 226-239.

JET DISPERSION IN CONFINED SPACES

Summary

This paper presents fundamental research work carried out on a submerged round jet discharging under three confined conditions: (i) oblique impingement in channels, (ii) vertical impingement in channels, and (iii) vertical dispersion into deep wells. The degradation and spreading of the boundaries of the jets are studied while varying the downstream depth of flow, width of channel, height and position of the jet, and the Froude number of the flow. For the vertical well, the ratios of the jet diameter well width and height are varied between the practical limits of 6 and 12. Generalised power laws are derived and compared with data on submerged jets impinging on a flat plate.

Nomenclature

a = Jet Height
C = Proportionally coefficient
d = Jet diameter in channel
D = Jet diameter in wall
F_o = Froude number of flow at jet outlet = $U_o/\sqrt{gd}$
h_1 = Height of impingement in channel
h_2 = Downstream channel depth
n = Exponent of power law
U_o = Jet exit velocity in channel/well
U_1 = Initial maximum velocity of deflected jet in channel
U = Maximum centreline velocity in channel
V = Maximum velocity in well
w = Width of channel/well
x = Distance along centreline of channel
y = Distance/height in well
ϕ = Angle of impingement

Introduction

Jet dispersion in an ambient fluid of infinite extent is a problem of fluid motion on which a considerable amount of experimental and theoretical investigation has been published, [1,2]. The more complex motion of a round jet discharging obliquely or vertically into a rectangular channel or deep pool however, has received far less attention. No records were found of fundamental work on the behaviour of such a jet or the effect of the depth and width of the pool or channel upon the dissipation of a jet entering it, a matter which obviously bears upon industrial design.

For many years, the author has been investigating the diffusion of turbulent confined jets and streams. Initially experiments were carried out to establish the shape of the velocity profiles, the decay of maximum velocity and the rate of growth of a round air jet projected parallel to a wall, [3]. It was found that in the outer half of the jet the rate of spreading parallel to the wall to be about 6 to 8 times greater than that normal to it. The research was extended by studying the flow characteristics of the jet projected obliquely to a plane smooth surface, [4]. It was discovered that at about 30° deflection angle the jet fans out radially on the surface. For the 30° and 45° deflected jets the decay of maximum velocity and head loss are much faster than those for other submerged jets; the decay rate being about two and a half times that for the parallel wall jet.

Experiments were then conducted on a jet issuing from a submerged horizontal pipe into an open rectangular channel. It was observed that the jet disperses with intense turbulence and asymmetry and is extremely unstable, [5]. For semi-submerged flow, the jet is sucked laterally and adheres to one side of the channel, with a long circulation zone forming on the other side. At high depth of submergence, the jet changes into a state of sustained movement from one side of the channel to another. Attempts to stabilise the jet without altering its character, were unsuccessful. However, reasonable steady and symmetrical flow was obtained when the jet was directed at about 45° and more to the channel bed. Immediately after issue the jet fanned out on the floor of the channel and ran up the side walls very smoothly, much of its excess energy dissipated at a short distance downstream.

The deceleration of the deflected stable jet was studied for various angles of impingement, [6,7]. The results showed that the maximum velocity is reduced to less than one quarter in a short distance of 8 diameters, which is equivalent to a dissipation of more than 85 per cent of the total head of the jet. Over the same distance the head decreases more than ten times faster as in the air jet in open spaces or a submerged jet dispersing in its own surrounding fluid.

The main aims of the present research were to carry out investigations into jet dispersion in rectangular channels and vertical wells. In the experiments the inclination of the jet was varied from 45° to 90° to the bed of the channel, while differing ratios of downstream depth to jet diameter were studied for various Froude numbers. The effects of the jet height above the bed, the width of the channel and jet distance from the upstream face of the channel were also explored. Vertical wells of square and rectangular cross sections were used in the tests. Measurements of horizontal and vertical velocity profiles were carried out using a Laser Doppler Anemometer. The results should have industrial applications in cutting, cleaning, mining and tunnelling, civil engineering and off-shore work.

Apparatus

The work was carried out in special water jet equipment consisting of a 25mm diameter pipe discharging with varying inclinations into a glass flume 1800mm long and 300mm wide. Perspex wall were inserted into the channel to vary its width. The flow was supplied by a centrifugal pump and measured by a digital meter. The depth of flow in the channel was controlled by means of an adjustable tailgate situated in the outlet tank.

The Laser Doppler Anemometry (LDA) technique was adapted for measurements of mean velocities and turbulence characteristics, without any interference in the flow and utilized for the investigation of the complex three dimensional circulating flow which was experienced in the channel. Figure 1 is a schematic diagram showing the main components of the LDA system, as well as the data acquisition and analysis systems.

Oblique Jet Impingement in Channels

A broad programme of research was at first drawn for the investigation of a round jet issuing at various angles into a channel, but following the completion of certain

experiments, it was considered that it would be more profitable to concentrate mainly on the case of the 45° oblique jet.

Flow Pattern

A schematic section of the motion of the oblique jet along the centreline of the channel is shown in Fig.2.

Immediately after issue the jet fans out on the floor of the channel. Between the fast stream and the surrounding fluid a turbulent mixing layer is formed having an outer boundary 02. The line 03 defines the mean flow line with the stream function $\psi = 0$, bounding the fluid which is sucked outward from the stream. The mixing process between the jet and the surrounding fluid causes a circulation zone to be established on top of the stream, in a similar manner to the flow behind a two-dimensional bluff body. The purpose of the investigation was to

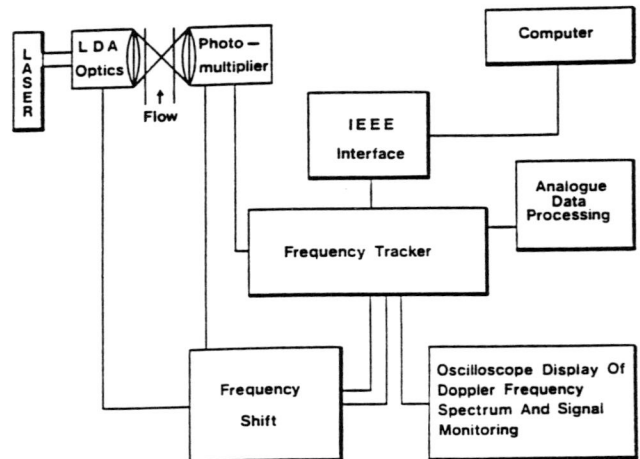

FIG.1. - MAIN COMPONENTS OF LDA SYSTEM

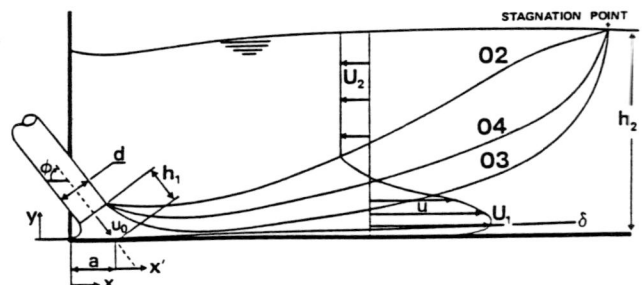

FIG.2. - OBLIQUE JET IMPINGEMENT IN A CHANNEL

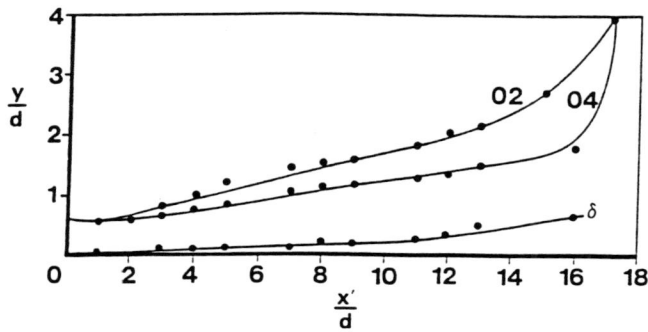

FIG.3. - JET BOUNDARIES FOR 45° IMPINGEMENT

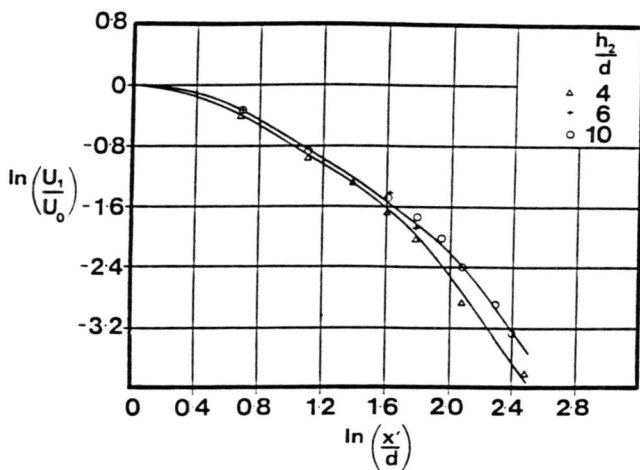

FIG.4. - DECAY OF MAXIMUM VELOCITY FOR VARIOUS DOWNSTREAM DEPTHS

Maximum Velocity Decay

The rate of decay of the maximum central velocity U_1 is plotted logarithmically in Figure 4. The flow is fully developed after a distance of 3d; thereafter its decay rate is of the form shown below:

$$\frac{U_1}{U_o} = C \left(\frac{d}{x^1} \right)^n$$

where C is a proportionality constant and n is an exponent.

For a downstream depth of 4d the values of these constants compare well with earlier work carried out by the authors [6 and 7]. In this region of flow the rate of decay is also similar to those found by Naib [4], Beltaos [9] and Yakovlevskii and Krasheninnicov [8] as shown in Table 1.

Following the entrainment region a transition takes place into the recirculation region where the jet decays much faster and resembles uniform channel flow shortly downstream of the end of the circulation zone. Seventy percent of the original velocity is lost within a distance of 8d and an additional 20% over the following 12d.

Decreasing the downstream depth changes the overall rate of decay. The maximum velocity remains inversely proportional to the distance, with n = 1.06 but the value of the proportionality constant C decreases thus indicating a shift in the virtual origin. This is the result of the presence of greater pressure gradients when the downstream water level is low, indicating a conversion of some of the kinetic head to static head.

In Figure 5 the non-dimensional plot of the velocity profiles for distances between 3d and 7d lie nearly on one curve and are therefore similar. The end of similarity of velocity profiles indicates the termination of the entrainment region and a change in the rate of decay.

Vertical Jet Impingement in Channels

Flow Pattern

The flow pattern illustrated in the definition sketch Figure 6 for vertical impingement clearly indicates a complex multiple-circulation region above the deflected jet.

It is apparent that the upper half of flow has been forced to move downstream instead of re-circulating as in the case of the 45° impingement. Observations of the free flow conditions (figure 7) show that on impact the deflected jet spreads along the bed and up the side walls of the channel.

The height of the stream on the side walls is dependent on

establish several parameters, including the rate of spreading, the decay of maximum velocity, the distribution and shape of the velocity profiles under different degrees of submergence. The results obtained are summarised below.

Jet Boundaries

Figure 3 shows the boundaries of the stream for a downstream depth of four jet diameters. The rate of growth of the boundary layer δ near the bed of the channel is linear with an inclination of 1°. The angles of spreading of the boundaries 02 and 04 are approximately 7° and 5.5° respectively.

The rate of growth of the normal half width $b(U_{(b)} = U_{1/2})$ is linear and follows the relationship $b = mx^1$. It was found that m=0.067 which corresponds to an angle of growth 3.8°. This was constant for a range of downstream depths between 4d and 8d. For an air jet impinging on a smooth flat surface at an angle of 45° Naib (4) suggest a value of m = 0.086 (4.9°) while the experiments of Yakovlevskii and Krasherinnicov (8) show b to grow linearly with a slope of 0.075 (4.3°).

the width of channel and the result of this is the flow being deflected and falling back into the centre of the channel. When the jet is submerged it is the fluid deflected from the side walls which causes the upper part of the circulation region to be forced into flowing downstream.

Jet Boundaries

The stream boundary 04 for different downstream depths is plotted in Figure 8. The mean angle of spreading of this boundary is 9°.

It is interesting to note that the stagnation point remains constant at about 4d for the range of depths tested. This is again the result of the flow deflected from the side walls. The geometry of the deflected jet is such that it remains the same for all depths of submergence, therefore

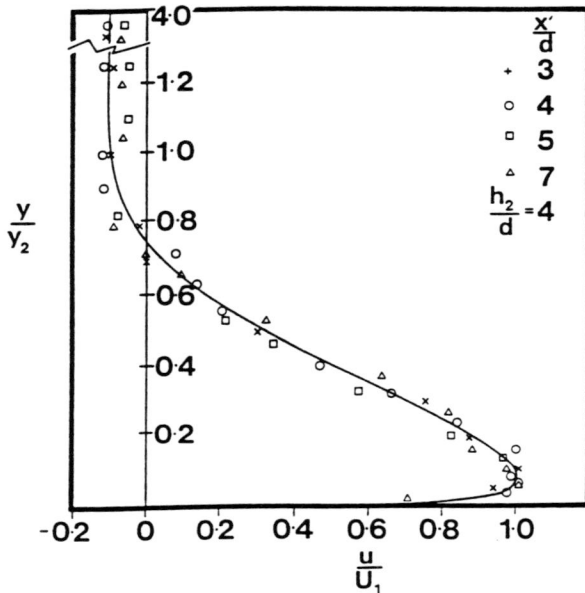

the flow from the side walls always falls back into the centre of the channel at the same position, hence the position of the stagnation point remains practically constant. The length of the submerged part of the circulation region is prevented from expanding normally to the surface by the flow from the side walls.

The rate of growth of the normal half width b for a depth of submergence of 10d, was found to be $b = 0.08x^1$ (4.6°). This is surprisingly in close agreement, with the result of an air jet impinging vertically on a flat plate [10, 11, 12].

Maximum Velocity Decay

Figure 9 gives examples of the distribution of maximum velocity along the centreline of the channel for various downstream depths. For the entrainment region $2d < x^1 < 5d$, the variations are linear and may be represented by the

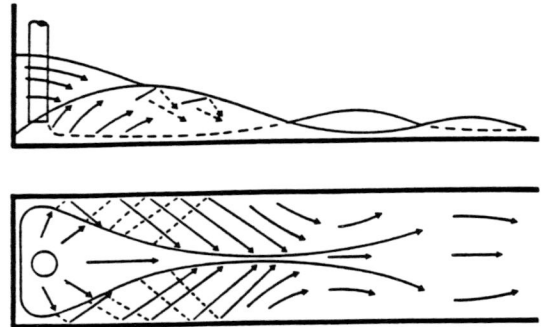

FIG.5. - NON-DIMENSIONAL PLOT OF VELOCITY PROFILES

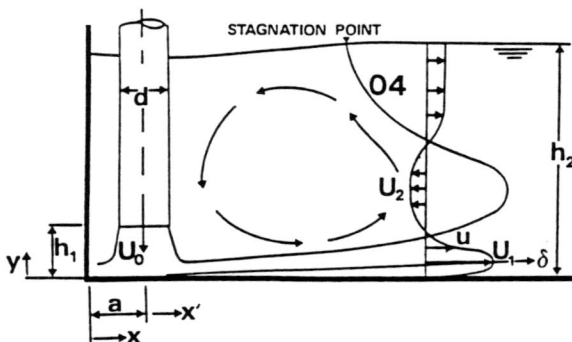

FIG. 7. - FREE FLOW CONDITIONS

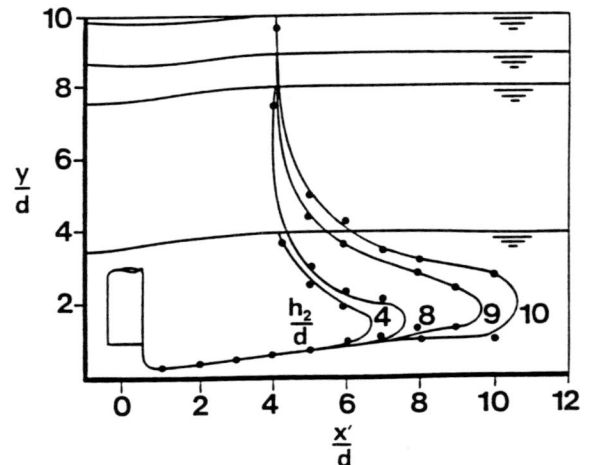

FIG.6. - VERTICAL JET IMPINGEMENT IN A CHANNEL

FIG. 8. - JET BOUNDARIES FOR 90° IMPINGEMENT

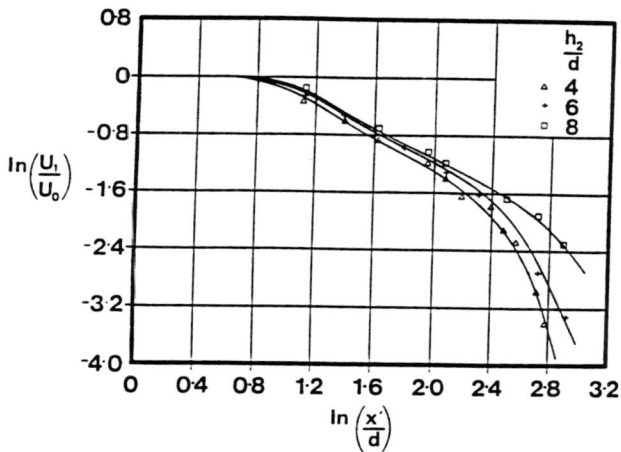

FIG.9. - DECAY OF MAXIMUM VELOCITY FOR DIFFERENT DOWNSTREAM DEPTHS

more accurate measurements, which were made possible due to the utilisation of the LDA equipment, show that after 5d the rate of decay changes. This may indicate the end of an entrainment region, and a transition into the recirculation region. The different rates of decay for varying depths may be due to the presence of a pressure gradient as evidenced by the slope of the water surface in the direction of flow. For depths greater than 6d the decay rate remains unchanged as the water surface becomes flatter at these depths and hence the change in pressure gradient is too small to have an effect on the flow.

Approximately 90 per cent of the pipe outlet velocity is lost over a distance 8d compared with only 70 percent for the 45° impingement.

Jet Position

The results of tests for various jet heights seem to indicate that increasing the height of impingement has little or no effect on the deceleration of the jet [13]. However, increasing the height of the jet causes the water level immediately upstream of the pipe outlet to rise and therefore for practical considerations it should be kept small. A height of one jet diameter above the channel bed is recommended. The positioning of the pipe from the back wall of the channel is another important factor and again from maximum velocity and minimum scour considerations a distance of one jet diameter from the back wall was found to give the least scour downstream.

following equations:

$$\frac{h_2}{d} = 4 \; ; \; \frac{U_1}{U_o} = 1.6 \left(\frac{d}{x'}\right)^{1.3}$$

$$\frac{h_2}{d} = 6, 8, 10 \; ; \; \frac{U_1}{U_o} = 1.7 \left(\frac{d}{x'}\right)^{1.3}$$

These results show some deviation from earlier work by the authors [6 and 7] where $U_1/U_o = 1.79(d/x^1)^{1.45}$ for $2 < (x^1/d) < 8$. This is thought to be due to the fact that in the earlier work carried out the decay was assumed to be at a constant rate between 2d and 8d. Whereas the current and

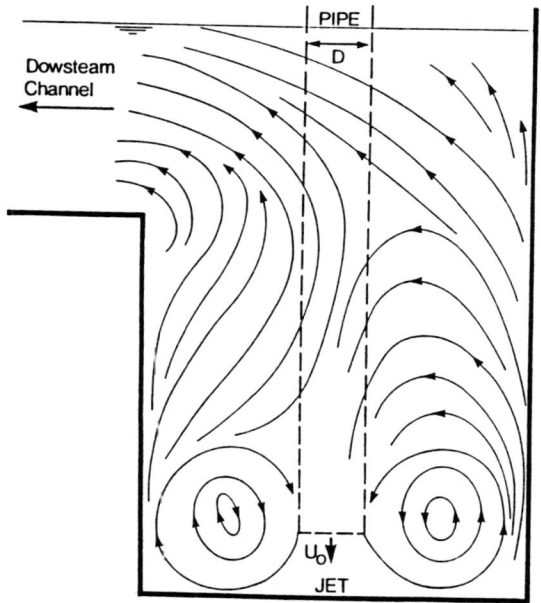

FIG. 10 - VERTICAL JET DISPERSION INTO A WELL

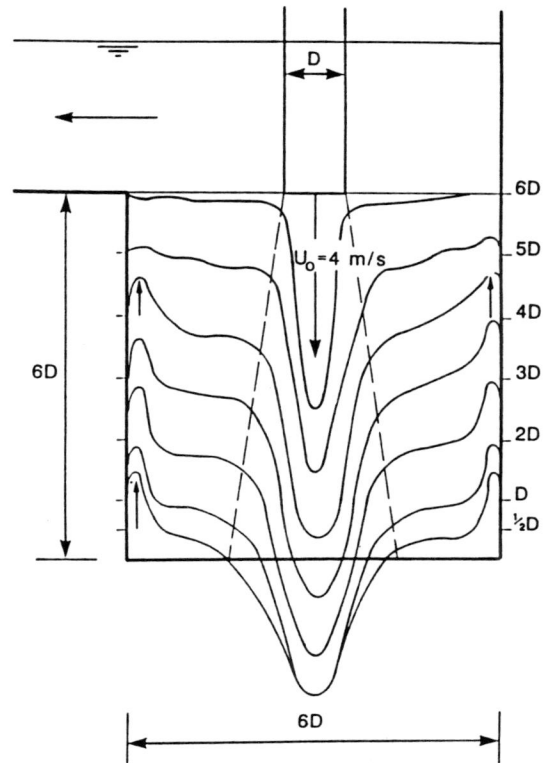

FIG. 11 - VELOCITY PROFILES AT VARIOUS HEIGHTS ABOVE THE FLOOR OF THE WELL

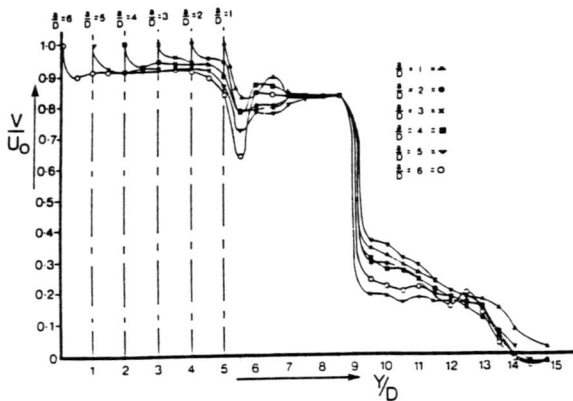

FIG. 12 - VARIATION OF MAXIMUM VELOCITY IN A RECTANGULAR WELL

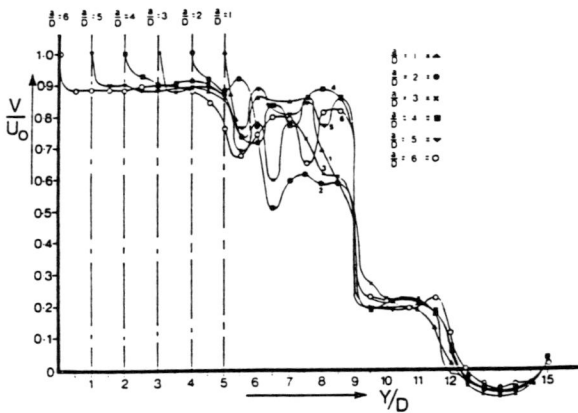

FIG. 13 - VARIATION OF MAXIMUM VELOCITY IN A SQUARE WELL

Froude Number

Experiments were also carried out for various Froude numbers and although the range was limited between 4 and 8, it seems fairly certain that there is no tendency for the rate of decay of the jet to change with Froude number. The corresponding Reynolds numbers were in the range of 10^4 to 10^5.

Vertical Jet Dispersion into a Well

Flow Pattern

When a round jet discharges vertically into a well, a complicated three-dimensional flow pattern is formed, which varies from section to section across the width of the channel. Despite this complexity, three regions may be generally identified along the centre plan as shown in Fig.10.

In Region I, the jet issues into a slow moving fluid and expands in a similar way to a submerged round jet in an infinite fluid. For jet heights of up to 5D above the floor of the well, this region resembles that of the potential core. The mixing process between the jet and the surrounding fluid causes part of the latter to be carried downward with the jet and regions of recirculation are thus established. In Region II, the jet impinges on the floor of the well and is transformed into a thin stream which flows radially and horizontally in all directions as in the case of a radial wall jet. When the stream reaches the side walls of the well, it is partly deflected upwards and partly laterally towards the four corners of the well. Strong currents flow along

TABLE 1

AUTHORS	C			n		
Naib/Sanders (present data)	h_2/d 4	6	8	$h_2/d=$ 4	6	8
	2.4	2.6	2.7		1.06	
Naib [4]	2.4			1		
Yakovlevskii and Krasheninncov [8]	2.3			1		
Naib/Sanders [5]	2.32			1.02		
Beltaos [9]	2.5			1		

80

the corners resulting in upsurges above the mean water level. At each corner the rising stream then deflects through 180° and plunges towards the centre of the well thus creating tremendous wave action and turbulence in the downstream channel. In Region III, the decaying jet flows upward along the vertical sides of the well, and then deflects horizontally towards the channel downstream. Complex recirculation regions are created inside the well. At about a third of the height of the well from the bottom, the jet seems to curve inwards into the well forming a strong recirculation region.

The flow pattern near the channel and well side walls has some resemblance to that along the centre-line but it differs in its recirculation zone near the top of the well.

Vertical Velocity Profiles

In order to study the details of the flow, distributions of mean vertical velocity were measured across the centre plane of the well at various heights above the floor of the well. Typical profiles for a square well of side 6D and depth 6D, non-dimensionalized by the maximum jet velocity U_o are plotted in Fig.11. The jet height is 6D and the depth in the channel is 3D.

At the jet outlet the distribution is characterised by approximately a constant high velocity, decreasing sharply at the boundary of the nozzle. The surrounding fluid moves with a slow velocity in the same direction. At 4 diameters above the bottom of the well, the maximum velocity decreases gradually to zero at about 2D from the centre of the jet and then reverses in direction to show the rise of the jet along the vertical sides of the well. Further downward, at 3D, 2D and 1D above the floor, the velocity profiles become flatter but the general shape remains unaltered. The profiles however were not found to be similar. The reduction in maximum velocity is nearly 20% over a depth of 5D. The corresponding rate of spreading of the jet boundary is estimated to be 12°.

Variation of Maximum Velocity

A series of experiments were carried out to measure the variation of the maximum velocity decay in the jet as it travels downwards from the pipe outlet along the centre line of the well, radially along the bottom of the well, and finally up the side wall of the well before it deflects to form the downstream channel flow, (Figs. 12 and 13).

The experiments were conducted for the following conditions:

(i)   various jet heights above the well floor ranging from one to six diameters.
(ii)   various shapes of the well, i.e.:
        (a) rectangular well breadth 6D, depth 6D and width 12D.

(b) square well breadth 6D, depth 6D and width 6D.

In each case the width of the channel was the same as that of the well, and the downstream depth was kept constant at 3D.

In Region I of the vertical downward flow, there is a reduction of 10 to 20% of the initial velocity of the jet. In Region II of horizontal radial flow a similar percentage of original velocity is lost. In Region III of upward flow along the sides of the well, the remaining 60-80% of the velocity is mostly dissipated. It can clearly be seen that the maximum velocity reduction occurs as the horizontal jet deflects vertically upward at the junction of the floor and sides of the well. By comparing the results for the rectangular and square wells, it can be deduced that the rate of dissipation of velocity is increased by decreasing the width of the well from 12D to 6D. However, raising the height of the jet above the bottom of the well has little effect on the decay rate and the tendency of increased height is towards maginally decreasing the impact velocity at the bottom of the well.

Conclusions

Oblique Jet Impingement in a Channel

(i)   The rate of decay of maximum velocity in the fully developed region is inversely proportional to a power n of the distance. The value of n varies from 1.02 for $\phi = 45°$ to 1.3 for $\phi = 90°$.
(ii)   The angle of spreading of the boundary 04 is about 5°, nearly the same as for an airjet impinging on a flat smooth plate.
(iii)   The maximum velocity is reduced to less than one quarter in a distance of 8 diameters.

Vertical Jet Impingement in a Channel

(i)   For maximum head dissipation, the jet should be positioned one diameter from the floor and back wall of the channel.
(ii)   Although the mean angle of spreading of the stream boundary 04 is 9°, the distance of the stagnation point remains constant at 4 diameters for all downstream depths.
(iii)   Raising the height of the jet up to 6 diameters above the channel bed has little effect on the decay rate.
(iv)   Over the range of 4 to 9 the Froude number has practically no effect on the flow pattern.
(v)   The dissipation of head is increased by decreasing the width of the channel.

Jet Dispersion in a Well

(i)   In the impingement region, the flow resembles that in the potential core of a round submerged jet.

(ii)   The horizontal velocity near the bottom edge of the well is practically the same for all jet heights tested.

(iii)   Greatest velocity reduction, up to 60% occurs as the jet changes from horizontal to vertical flow along the four sides of the well.

(iv)   The radial flow along the floor of the well is primarily diverted into the four corners, resulting in strong currents and upsurges above the water surface. The plunging back of these jets into the centre of the well, leads to well known turbulence and surface waves in the downstream channel.

References

1.   Abramovich, G N "The Theory of Turbulent Jets" The MIT Press 1963.

2.   Rouse, B "Advanced Mechanics of Fluids" John Wiley & Sons, 1965.

3.   Naib, S K A "Spreading and Development of the Parallel Wall Jet" Aircraft Eng. pp 30-33, December 1969.

4.   Naib, S K A "The Deflexion of a Submerged Round Jet to Increase Lateral Spreading" La Houille Blanche, Vol.29, No.6, pp 455-461 October 1974.

5.   Naib, S K A and Sanders, J E "Diffusion of Bluff Wall Jets in Finite Depth Tailwater". Discussion paper 18354 - HY, ASCE, J Hyd. Div., Feb 1985.

6.   Naib, S K A and Sanders, J E "Jet Dispersion Downstream of Pipe Outlets" BHRA International Conference on the Hydraulics of Pumping Stations, Proceedings, Paper 10, Manchester, England 17-19 September 1985.

7.   Naib, S K A and Snaders J E "Jet Dispersion in Channels" Proc. Int. Conf. on Theoret. & Appl. Mech., University of South Carolina, USA, April 1986.

8.   Yakovlevskii, O V and Krasheninnicov, S Y "Spreading of a Turbulent Jet Impinging on a Flat Surface" Fluid Dynamics Vol.1, No.4, pp 136-139 July-August 1966.

9.   Beltaos, S "Oblique Impingement of Circular Turbulent Jets" Journal of Hydraulic Research, Int. Ass., for Hydraulic Research, Vol.14., No.1, pp 17-36 1976.

10.   Bradshaw, P and Lowe, E M "The Normal Impingment of a Circular Air Jet on a Flat Surface", ARC, R and M, No.3205, 1959.

11.   Poreh, M. Tsuei, Y G and Cernak, J E "Investigations of a Turbulent Radial Wall Jet" J. of Applied Mechanics, June 1967, pp 457-463.

12.   Erycake, P. Lee, D T, Gauntner, J W and Livingood J N B "Experimental Flow Characteristics of a Single Turbulent Jet Impinging on a Flat Plate" US NASA, TN-D-5690, 1970.

13.   Rajaratnam, N "Development in Water Science; Turbulent Jets", Elsevier, pp 226-239, 1976.

PART VI

JET DISPERSION IN POOLS AND CHAMBERS

Jet Dissipation in Deep Pools
Jets in Pipes and Chambers

JET DISSIPATION IN DEEP POOLS

Abstract

This paper presents an experimental investigation of a submerged round jet discharging vertically into a deep well. The flow pattern and decay of the jet are studied whilst varying the ratios of jet diameter to well width and height between the practical limits of 6 and 12. The downstream channel depth and Froude number of the flow are kept constant.

Nomenclature

a =   Jet Height
D =   Jet Diameter
U_o =   Jet Exit Velocity
U =   Vertical Velocity in Well
u =   Velocity in Channel
V =   Decaying Velocity in Well
X =   Distance Along Channel
Y =   Distance / Height in Well

Introduction

The dispersion of a free jet in a stationary surrounding has been extensively investigated and a considerable amount of experimental and theoretical investigation has been published, (Abramovitch 1963 and Schlichting 1979). The more complex motion of a round water jet discharging obliquely or vertically into a channel or into deep pools however, has received far less attention. Little records were found of fundamental work on the behaviour of such a jet or the effect of the depth and width of the pool or channel upon the dissipation of a jet entering it, a matter which obviously bears upon industrial design.

Vertical stilling wells and deep pools are normally used for high velocity discharge from pipe outlets. The pipe enters the well vertically, and in some cases, with a control valve attached to the end of the pipe.

The jet spreads horizontally along the bottom of the well and then rises vertically where it is dispersed and discharged with a low velocity into the downstream channel.

For many years, the author has been investigating the diffusion of turbulent confined jets and streams and some of the results have been reported elsewhere (Naib 1969 to 1991). The main aims of the present research were to carry out investigations into jet dissipation in deep pools and vertical wells in order to determine an efficient form of stilling basin for high velocity discharge from a pipe outlet and the minimum size of the basin for any given flow conditions. Wells of square and rectangular cross sections were used in the tests. Measurements of horizontal and vertical velocity profiles were carried out using a Laser Doppler Anemometer. Some of the preliminary results are given in this paper. These should have industrial applications in cutting, cleaning, mining, tunnelling, civil engineering and offshore works.

Apparatus

The work was carried out in a special water jet equipment consisting of a 25mm diameter pipe discharging into a glass flume 1800mm long and 300mm wide. Perspex walls were inserted into the channel to form the well. The flow was supplied by a centrifugal pump and measured by a digital meter. The depth of flow in the channel was controlled by means of an adjustable tailgate situated in the outlet tank.

The Laser Doppler Anemometry (LDA) technique was adapted for measurements of mean velocities and turbulence characteristics, without any interference in the flow and utilized for the investigation of the complex three dimensional circulating flow which was experienced in the well. Figure 1 is a schematic diagram showing the main components of the LDA system, as well as the data acquisition and analysis systems.

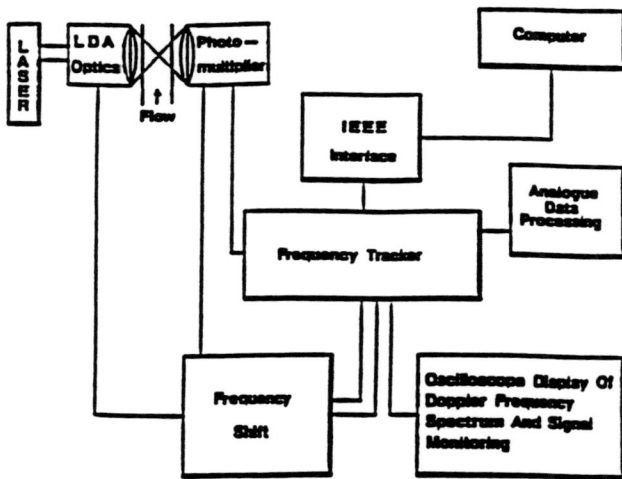

Fig.1 - Main Components of LDA system

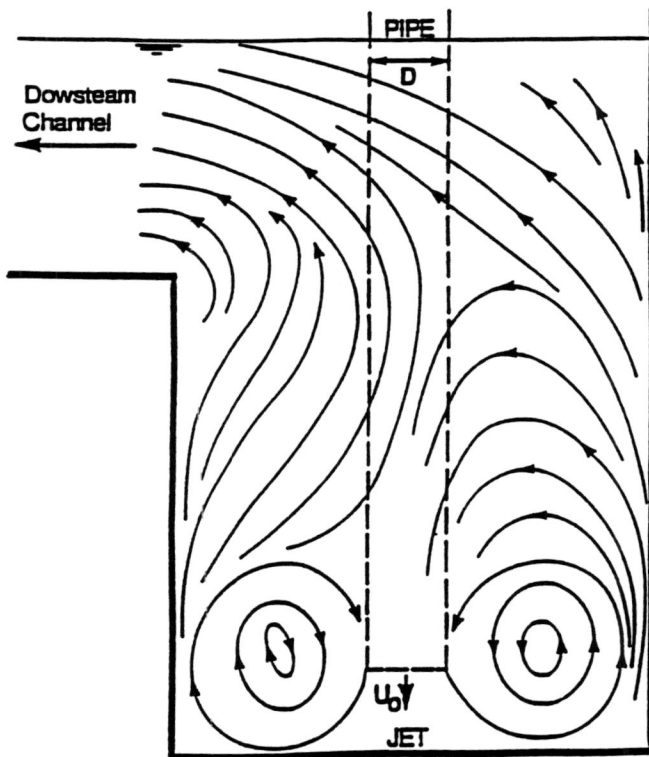

Fig.2 - Vertical Jet Dispersion into a Well

Flow Pattern

When a round jet discharges vertically into a well, a complicated three-dimensional flow pattern is formed, which varies from section to section across the width of the pool. Despite this complexity, three regions may be generally identified along the centre plane as shown in Fig.2.

In Region I, the jet issues into a slow moving fluid and expands in a similar way to a submerged round jet in an infinite fluid. For jet heights of up to 5d above the floor of the well, this region resembles that of the potential core. The mixing process between the jet and the surrounding fluid causes part of the latter to be carried downward with the jet and regions of recirculation are thus established.

In Region II, the jet impinges on the floor of the well and is transformed into a thin stream which flows horizontally in all directions as in the case of radial wall jet. When the stream reaches the side walls of the pool, it is partly deflected upwards and partly laterally towards the four corners of the well. Strong currents flow along the corners resulting in upsurges above the mean water level. At each corner the rising stream then deflects through 180 degrees and plunges towards the centre of the pool thus creating tremendous wave action and turbulence in the downstream channel.

In Region III, the decaying jet flows upward along the vertical sides of the well, and then deflects horizontally towards the channel downstream. Complex recirculation regions are created inside the well. At about a third of the height of the well from the bottom, the jet seems to curve inwards into the well forming a strong recirculation region. The flow patterns near the sides of the channel and pool have some resemblance to that along the centre-line but it differs in its recirculation zone near the top of the well.

Distribution of Vertical Velocity

In order to study the details of the flow, distribution of mean vertical velocity was measured across the central plane of the well at various heights above the floor of the well. Typical profiles for a square well of side 6D and depth 6D, non-dimensionalized by the maximum jet velocity U_o are plotted in Fig.3. The jet height is 6D and the depth in the channel is 3D.

At the jet outlet the distribution is characterised by approximately a constant high velocity, decreasing sharply at the boundary of the nozzle. The surrounding fluid moves with a slow velocity in the same direction. At four diameters above the bottom of the well, the maximum velocity decreases gradually to zero at about 2D from the centre of the jet and then reverses in direction to show the rise in the jet along the vertical sides of the well. Further

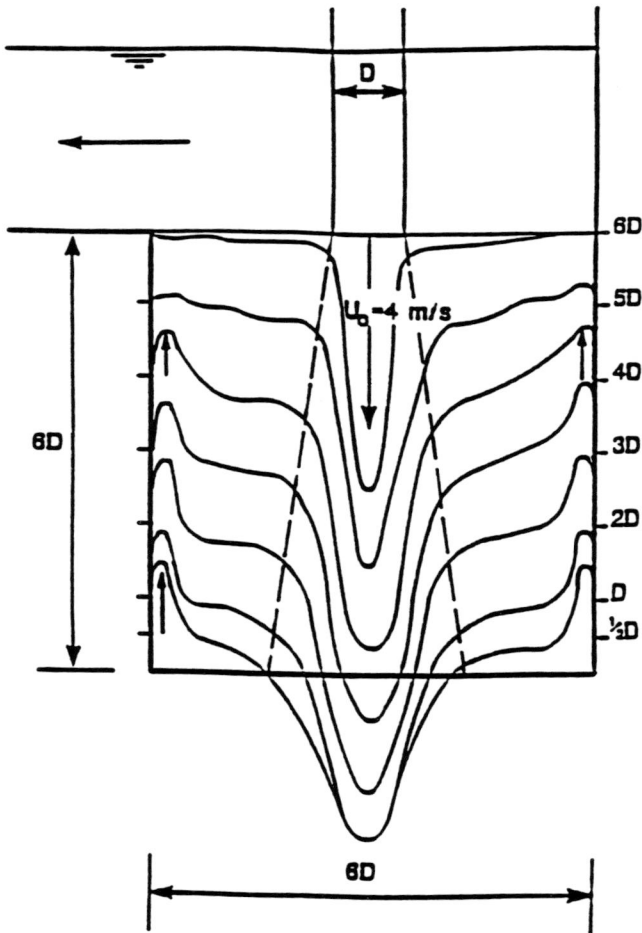

Figure 3 : Velocity Profiles at Various Heights Above Floor of Well

downward, at 3D, 2D and 1D, above the floor, the velocity profiles become flatter but the general shape remains unaltered. The profiles however, were not found to be similar. The reduction in maximum velocity is nearly 20% over a depth of 5D. The corresponding rate of spreading of the jet boundary is estimated to be 12 degrees, which is the same as that of a free round jet discharging in to the atmosphere.

Decay of Maximum Jet Velocity

A series of experiments were carried out to measure the variation of the maximum velocity decay in the jet as it travels downwards from the pipe outlet along the centre line of the well, radially along the bottom of the well, and finally up the side wall of the well before it deflects to form the downstream channel flow.

The experiments were conducted for the following conditions:

(i)   various jet heights above the well floor ranging from one to six diameters.

(ii) various shapes of the well, i.e.:
      (a) rectangular well breadth 6D, depth 6D and width 12D.
      (b) square well breadth 6D, depth 6D and width 6D.

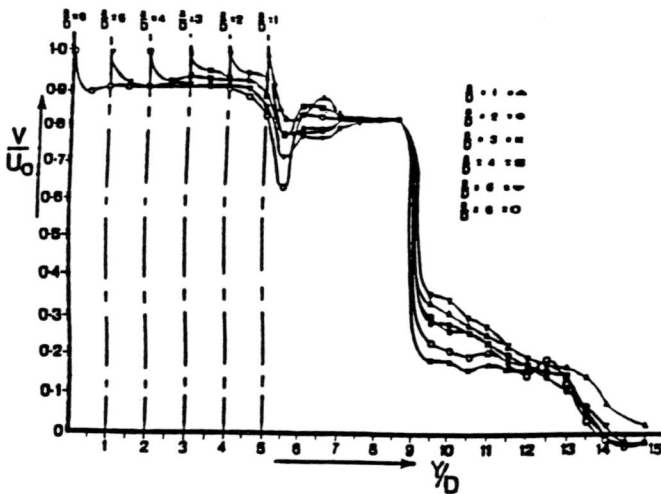

Figure 4 : Variation of Maximum Velocity in a Rectangular Well

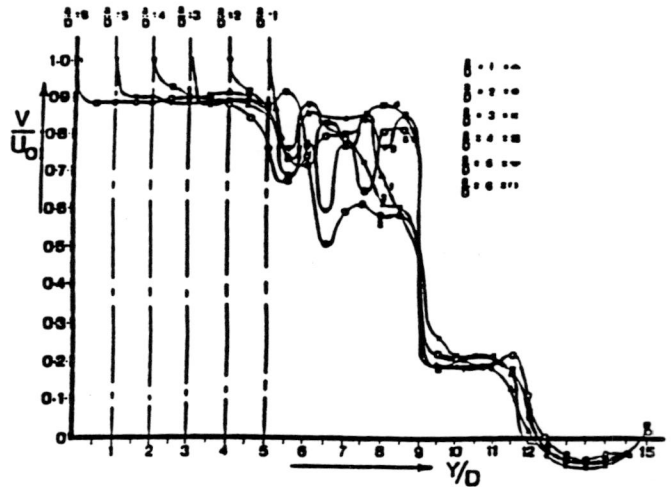

Figure 5 : Variation of Maximum Velocity in a Square Well

86

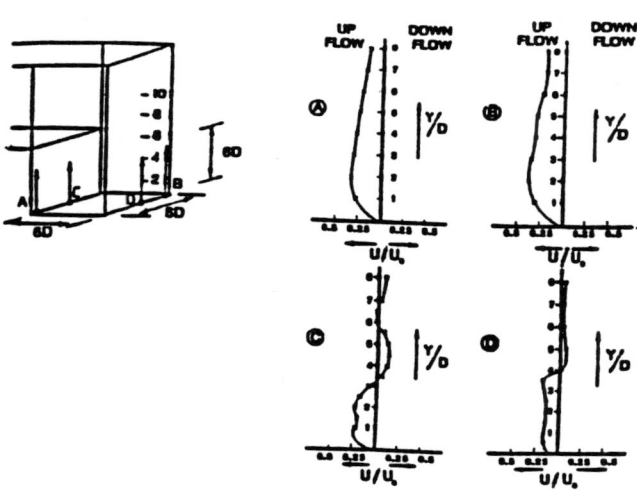

In each case the width of the channel was the same as that of the well and the downstream depth was kept constant at 3D. The variations of the maximum jet velocities are shown in Figs 4 and 5.

In Region I of the vertical downward flow, there is a reduction of 10 to 20% of the initial velocity of the jet. In Region II of horizontal radial flow a similar percentage of original velocity is lost. In Region III of upward flow along the sides of the well, the remaining 60-80% of the velocity is mostly dissipated.

Fig.6 - Vertical Velocity Decay Along Corners and Centre Line of Well

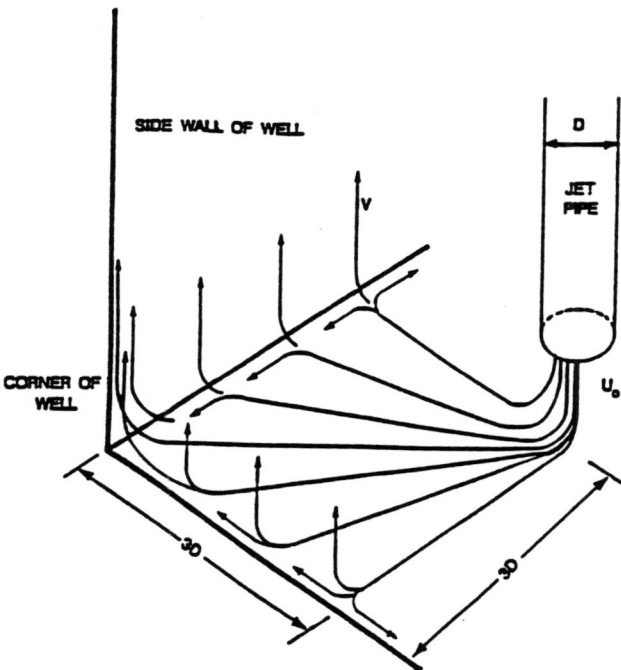

Fig.7 - Flow Pattern Along one Quarter of the Floor of the Well Towards a Corner

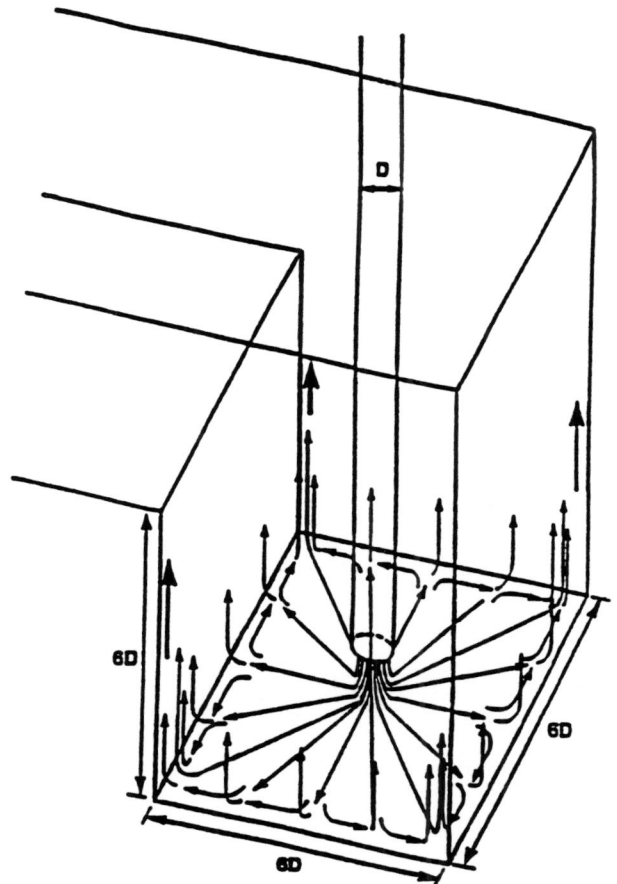

Fig.8 - Flow along the floor and sides of the well

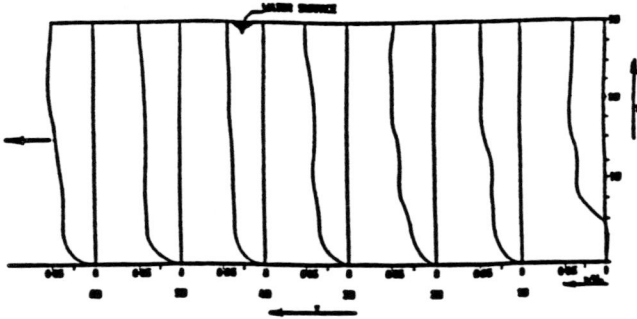

Figure 9 : Vertical Velocity Distribution in Channel Downstream of Well

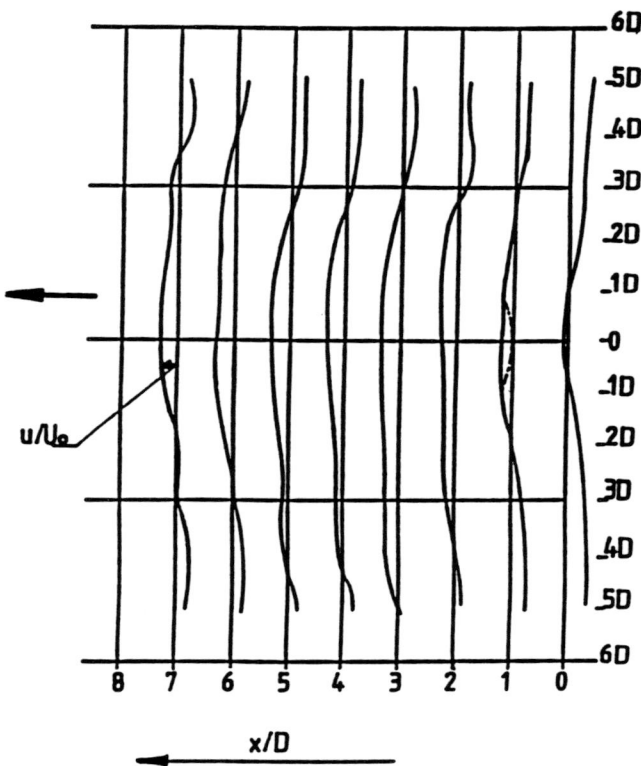

Figure 10 : Lateral Velocity Distribution in Channel Downstream of Well

It can be clearly seen that the maximum velocity reduction occurs as the horizontal jet deflects vertically upward at the junction of the floor and sides of the well. By comparing the results for the rectangular and square wells, it can be deduced that the rate of dissipation of velocity is increased by decreasing the width of the well from 12D to 6D. However, raising the height of the jet above the bottom of the well has little effect on the decay rate and the tendency of increased height is towards marginally decreasing the impact velocity at the bottom of the well.

Corner Velocity Profiles in Well

Measurements of vertical velocity in two corners of the well A and B and at two points C and D along the centre line are shown in Fig.6. At the corners all the flows are upwards and the strong currents rise to the water surface and plunge back into the pool creating water "boils" as high as 1.5D above the water line. In contrast, at point C and D the stream rises roughly to above 3D from the floor and recirculates in the lower half of the well. The corner currents form a second recirculation region in the upper half of the well. Figures 7 and 8 show the path of the jet and its spreading along the floor of the well.

Vertical Velocity Distribution in Channel

The velocity profile along the centre line of the channel downstream of a rectangular well of 12 diameter width were measured, for six stations at one diameter intervals, to study the transformation of the flow. Their profiles are plotted in Fig.9.

At the junction of the well and the channel there is a small recirculation zone near the channel bed. The velocities are in the reversed direction up to half a diameter height. The flow keeps fluctuating in this region in both forward and backward direction, and a small reversed flow is the mean indication in the LDA equipment for 5000 samples analyzed. Higher up from 0.5 to 3 diameter the velocity ratio (u/U_o) is about 0.04. Hence the forward velocity in the channel is approximately 4% of the original velocity.

Further downstream the velocity profiles are gradually transformed to resemble that of an open channel flow, remaining practically constant at 5% of the original flow from the channel bed to the water surface. The slight increase in the forward velocity at 6D is possibly due to the flow from the side wall being redistributed to the centre of the channel.

88

Lateral Velocity Distribution in Channel

The lateral velocity profiles near the bed of the channel plotted non-dimensionally in Fig.10, indicate forward flow in the centre of the channel and reverse flow at the side walls. With distance downstream the forward flow at the centre line of the channel gradually engulfs the centre half of the channel. It was noted the forward flow in the centre of the channel did not remain in the same plane all the time but oscillated form one side of the channel to the other quite frequently. The maximum non-dimensional forward velocity and reverse velocities were 0.05 and 0.06 respectively.

Conclusions

i   When a water jet discharges vertically into a well, a complicated three dimensional flow pattern occurs.

ii   In the impingement region, the flow resembles that in the potential core of a round submerged jet.

iii   The horizontal velocity near the bottom edge of the well is practically the same for all jet heights tested.

iv   Greatest velocity reduction, up to 60% occurs as the jet changes from horizontal to vertical flow along the four sides of the well.

v   The forward velocity in the downstream channel is about 5% of the exit jet velocity.

vi   The radial flow along the floor of the well is primarily diverted into the four corners, resulting in strong currents and upsurges above the water surface. The plunging back of these jets into the centre of the well, leads to the well known turbulence and surface waves in the downstream channel.

References

1.   Abramovich, G N. 1963. "The Theory of Turbulent Jets". The MIT Press.

2.   Naib, S K A. 1969. "Spreading and Development of the Parallel Wall Jet" Aircraft Eng, p.30.

3.   Naib, S K A. 1974. "The Deflexion of a Submerged Round Jet to Increase Lateral Spreading" La Houille Blanche, Vol 29, No 6, pp 455-461.

4.   Naib, S K A. and Sanders, J E. 1985a. "Diffusion of Bluff Wall Jets in Finite Depth Tailwater". Discussion paper 18354 - HY ASCE, J Hyd. Div.

5.   Naib, S K A and Sanders, J E. 1985b. "Jet Dispersion Downstream of Pipe Outlets". BHRA International Conference on the Hydraulics of Pumping Stations, Proceedings, paper 10, Manchester, England.

6.   Naib, S K A and Sanders, J E, 1986. "Jet Dispersion in Channels". Proc. Int. Conf. on Theoret. & Appl. Mech., University of South Carolina, USA.

7.   Naib, S K A, Sanders, J E & Rasiah, V. 1988. "Oblique and Vertical Jet Dispersion in Confined Spaces". Proc. of 9th Int. Conf. on Jet Cutting Technology, Sendai, Japan.

8.   Naib, S K A. 1991. "Jet Mechanics and Hydraulic Structures". Forthcoming Book.

9.   Schlichting, H. 1979. "Boundary Layer Theory". McGraw Hill Book Co.

JETS IN PIPES AND CHAMBERS

Abstract

In desalination and other engineering work, it is often required to discharge water from a pipe at high pressures into an outfall culvert at atmospheric pressure. A combination of pressure reducing orifices or valves in conjunction with high velocity energy dissipators in the form of a deep pool are normally used to achieve this requirement. This paper describes fundamental research work on jet dissipation in pipes, closed chambers and deep stilling wells in order to study the details of flow pattern, velocity distribution, and energy reduction. The effects of chamber dimensions and jet heights are also investigated.

Nonmenclature

a	=	Jet height
B	=	Chamber width/pipe diameter
D	=	Jet diameter in well/pipe
h_2	=	Downstream channel depth
U_o	=	Jet exit velocity in pipe/well
U	=	Maximum centreline velocity in pipe
V	=	Maximum velocity in well
x	=	Distance along centreline of pipe/well
Y	=	Distance/height in well

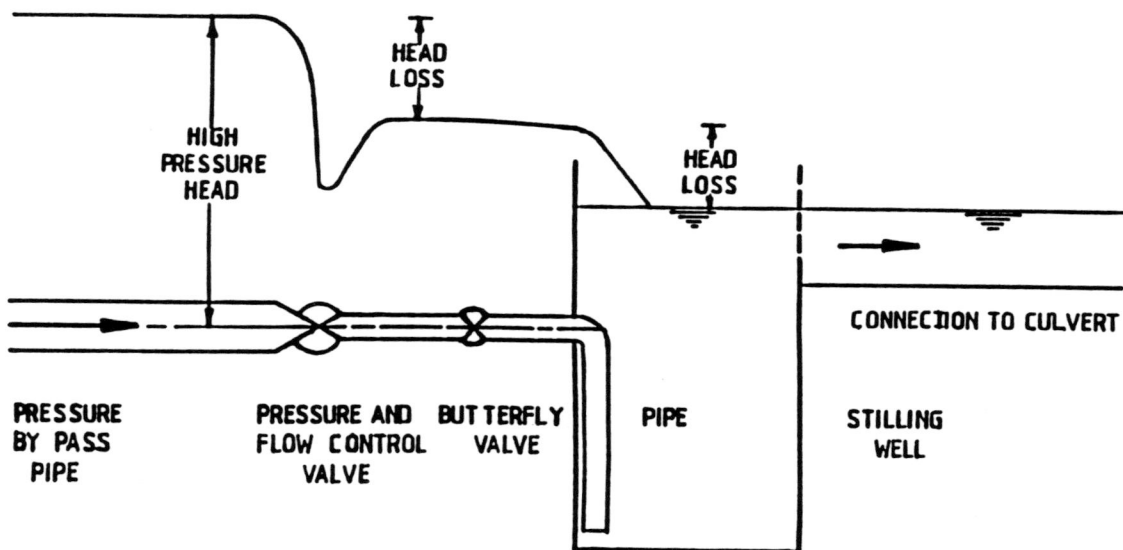

Fig.1 - Pressure reductions in pipelines

Introduction

In the design of desalination and power plants, pressure pipe bypasses have to be provided in the system with connections to outfall culverts. Because of the specification requirements and the restrictions brought about by the use of fixed speed pumps, dumping through the bypasses occurs as part of normal plant operations. The main problems associated with the design are the dissipation of the pressure head between the last control valve and the junction of the outfall culvert and the prevention of unacceptable local levels of swell with serious consequences in the operation of the culvert. The practical solution is to provide some form of energy dissipation between the last valve and connection to the outfall culverts. In all cases model studies are carried out to verify the design and predict the performance of the prototype.

Figure 1 shows a possible system which consists of three parts: (1) a pressure and flow control valve which maintains a specified pressure by dumping the excess flow; (2) a fixed orifice jet which has a constant flow/head loss characteristic and provides sufficient pressure downstream of the pressure and flow control valve to prevent the formation of low pressures and hence cavitation; (3) a deep stilling well which dissipates the energy of the jet and stabilizes the flow before it enters the outfall culvert. The principle of the pressure reducing valve is to convert the pressure head into a velocity head which is then dissipated by turbulence downstream of the valve. Alternatively, energy is dissipated by spreading the water through a nozzle into the air, Fig.2. The disperser nozzle has a number of deflector vanes producing a vortex motion which causes the jet to disperse over a wide angle. The jet is designed to disperse the water into a spray so that it falls as heavy rain on a pool with a concrete apron. The energy is dissipated by the resistance of the droplets into air. In this research only investigations of the dispersions of the jets in the pipeline and the stilling well have been carried out to determine the pattern of flow and decay of the jets. Some of the preliminary results are presented in the following sections.

Apparatus

The experimental equipment consisted of 8, 16, 25 and 50mm diameter pipes discharging into a perspex pipe or a well fitted in a glass flume 1800mm long and 300mm wide. Perspex walls were inserted into the channel to vary the width and depth of the well. The flow was supplied by a centrifugal pump and measured by a digital meter. The Laser Doppler Anemometry (LDA) technique was adapted for measurements of mean velocities and turbulence characteristics. The apparatus for the air jet experiments consisted of a centrifugal fan fitted with a 20mm nozzle and a perspex pipe. For measuring the velocity a total head Pitot tube was used.

Jet Dispersion in Pipes and Chambers

Experiments were carried out with air and water jets being discharged into a pipe. The ratio of pipe to nozzle diameter (B/D) was varied between 3 and 10.
The axial maximum velocity and the lateral distribution of velocity across the pipe (Fig.3) were measured up to a distance of 20 diameters downsteam. Figure 4 shows a non-dimensional plot of the maximum velocity of the jet along the centreline of the pipe. It is seen that the decay of velocity is most marked when the ratio of pipe to jet diameter is three. For this condition the velocity of the jet decreases much faster in the confined jet than in a free air jet discharging into the atmosphere, shown by the lowest curve.

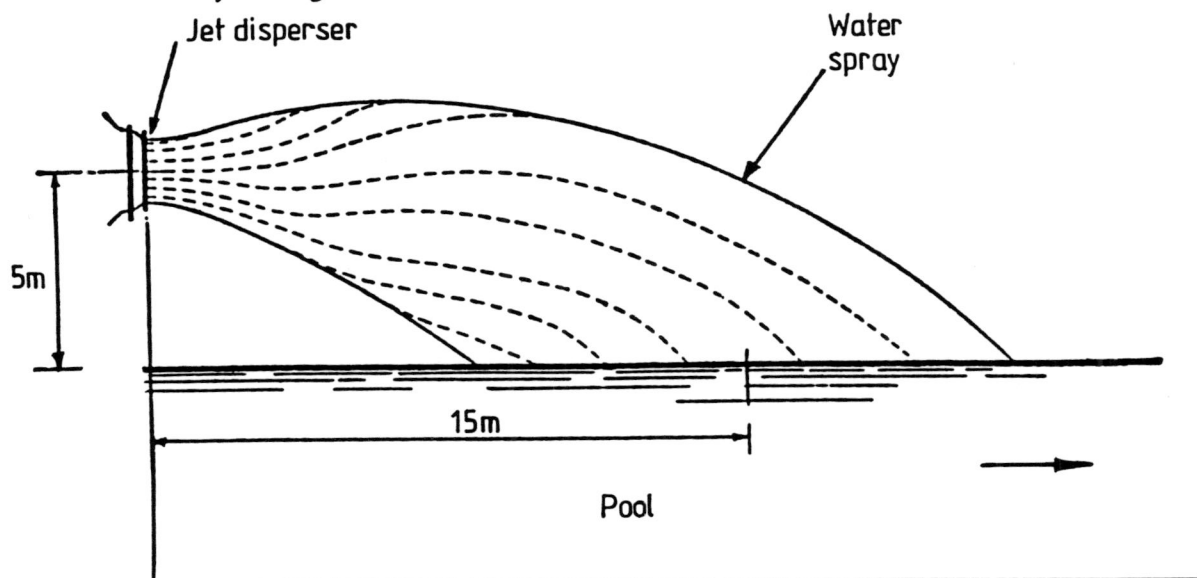

Fig 2 - Diffused discharge from a jet disperser

Following potential core of about 4 diameters, the velocity in the pipe decreases twice as fast (and the kinetic head four times as fast) compared to the open jet. At a distance x/D = 12 to 14 the velocity falls off five times faster than in the free jet. For larger ratios of pipe to jet diameter of 6 to 10 the decay of velocity follows the curve for the open jet. An increase, therefore, in the diameter of the pipe does not improve the rate of dissipation along the centreline. Similar experiments with an air jet discharging into a circular pipe showed the same trend and that the maximum velocity is reduced considerably at a distance of 20 diameters from the jet orifice. Other results are described below:-

1 The length of the potential core is about four diameters from the jet outlet, compared with five diameters for a free jet.
2 Near the orifice in the transition region, the velocity decrease twice as fast in the pipe as in the open air jet.
3 In the fully established flow, $8 < x/D < 20$, the effect of the pipe/chamber width β in decreasing the maximum jet velocity is most marked when its diameter is two to three times that of the jet.
4 For $B/D > 6$, the maximum velocity is reduced to less than one quarter in a distance of 12D which is equivalent to a dissipation of more than 90% of the original kinetic head of the jet.
5 For $B/D > 6$, the width of the chamber has little effect and the maximum velocity follows the decay curve of the open jet. The tendency of increased width is towards increasing the velocity at the core of the jet.
6 For $B/D = 3$, the velocity id decreased to about one tenth of the original jet velocity in a distance of 20 diameters. At this distance the jet has fully diffused across the width of the chamber.
7 For all widths of chambers and jet velocities, air pockets with free water surface are formed below the top of the chamber, extending for a number of diameters downstream.

Applying these results to the practical design of pressure reducing valves, the indications are that the dissipation is greatest when the width/diameter of the pipe is about 3 times the diameter of the jet and that the jet disperses over a distance of 20 diameters, with the velocity decreasing to one tenth of the original value. A dissipation of more than 90% of the energy of the jet will ensure smooth flow in

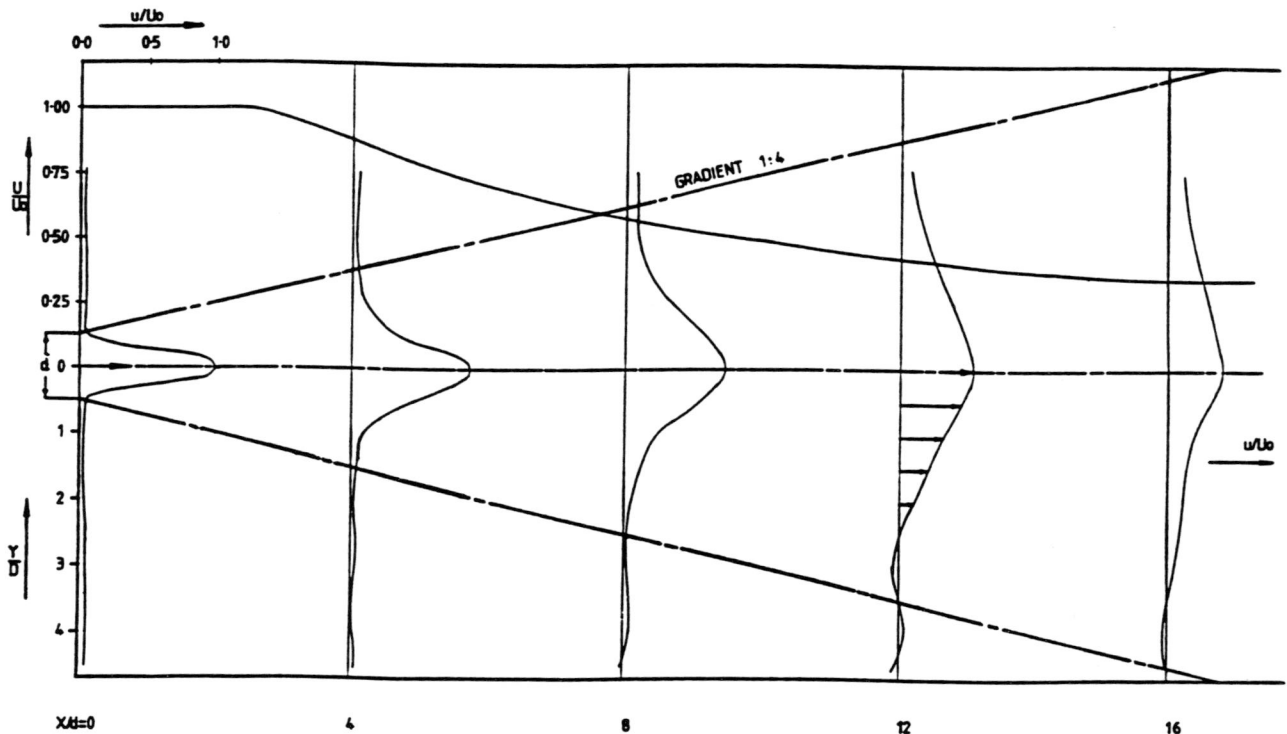

Fig 3 - Velocity distributions for jets in pipes

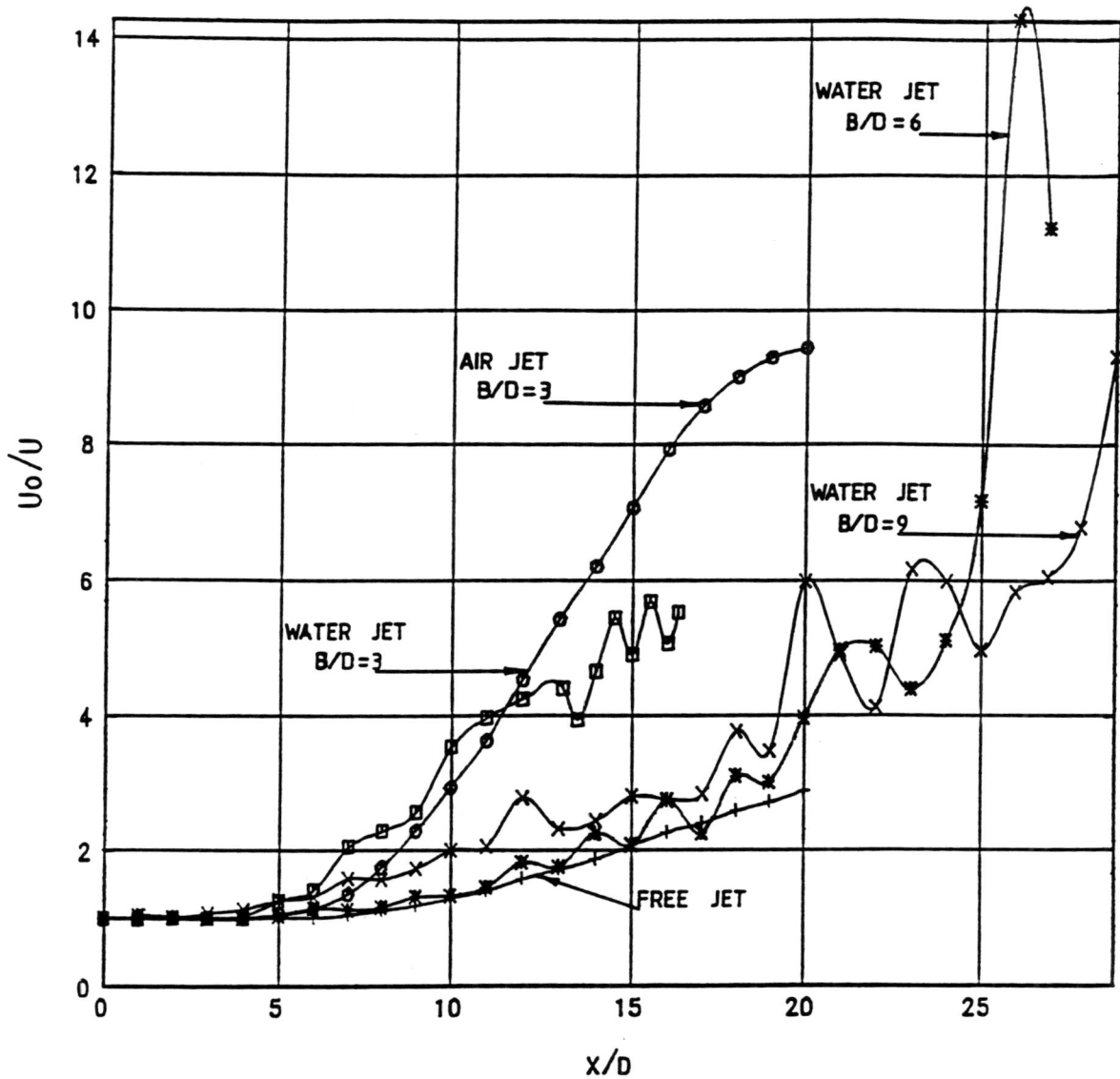

Fig 4 - Jet dissipation in pipes

the pipe downstream of the valve. This will require a minimum length of pipe downstream of the orifice of 12 times and a width of 3 times the diameter of the orifice. For a 600 diameter pipe, the orifice size should be 200mm and the length of the pipe downstream should be at least 2.4m.

Jet Dissipation into Deep Wells

When a round jet discharges vertically into a well, a complicated three-dimensional flow pattern is formed, which varies from section to section across the width of the well. A series of experiments were carried out to measure the variation of the maximum velocity decay in the centre line of the well, radially along the bottom of the well, and finally up the side wall of the well before it deflects to form the downstream channel flow. The experiments were conducted in a square well of 6D side for various jet heights above the well floor ranging from 1 to 18D. In each case the width of the channel was the same as that of the well and the downstream depth which was kept constant at 3D. Tests in a well of 4D side produced substantial surface waves in the downstream channel. The results are compared in FIg.5, together with the decays of the open jet and the pipe jet. Generally, for all heights the deceleration of the jet in the well follows that of the open jet. For deep wells the rate of dissipation increases as the jet reaches the bed of the well. The results are readily available for design purposes.

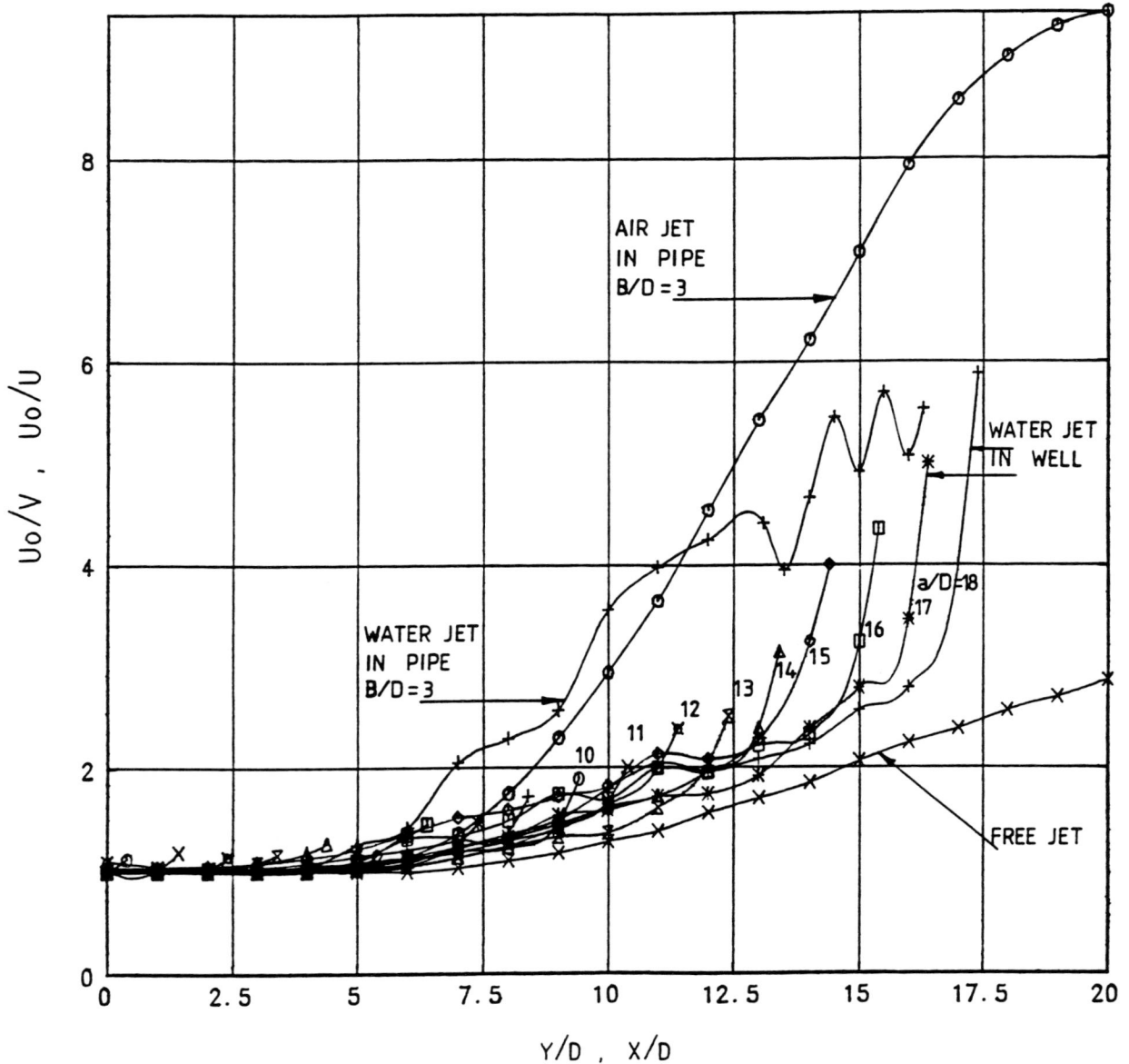

Fig.5 - Jet dissipation in deep wells

94

PART VII

JET DIFFUSION DOWNSTREAM REGULATOR STRUCTURES

Flow Under a Sluice Gate
Unsteadiness of Circulation Pattern
Abnormal Grouping of Large Eddies
Photographic Method for Measuring Velocities
Diffusions of the Sluice Way Jet

FLOW UNDER A SLUICE GATE

Introduction

When a stream passes under a submerged sluice gate into water on the downstream side, it forms a jet which goes far downstream before it is dispersed. The physical aspects of the dispersion of such a jet have some similarity, both qualitatively and quantitatively, with the simpler problem of the diffusion of fluid jets[1]. However, the flow of an air jet discharging from a slit into the atmosphere is not affected by adjacent boundary restrictions. A sluice-way jet, on the other hand, disperses into restricted surrounding fluid and has several fundamental properties which distinguish it from a free air jet. Apart from the contraction effect due to the sharp edge of the gate, the differences are as follows.

The jet issues into a slow-moving medium and is bounded on one side by a solid boundary and on the other side by a free-water surface. The mixing process between the jet and the surrounding fluid causes part of the latter to be carried forward with the jet under conditions in which both the forward momentum and discharge are conserved. This entrainment leads to recirculation to replace the fluid entrained; thus a zone of circulation is established on top of the jet which is usually termed a 'roller'. Meanwhile, the mixing process causes the jet to expand laterally until its upper mean flow line reaches the free surface. This point is the stagnation point which defines the downstream end of the circulation zone.

In comparision with the free jet, the chief difference lies in the presence of the circulation zone. There is also the difference that in the sluice jet the upper stream or limit of the flow is a surface which is free from shear stress and slopes upward in the downstream direction, indicating that the diffusion process takes place under a pressure gradient, often of considerable magnitude.

Experimental Method

The rise of pressure in the direction of flow is associated with the conversion of some of the kinetic energy of the high-velocity jet into pressure. The resulting difference of pressure influences the discharge through the sluiceway and also influences the nature of the flow patterns downstream as indicated in Figs.1 and 2. The former concerns variation in downstream depth with the Froude number of the flow $F_1 = q/(gd_1^3)^{0.5}$, where q is the unit discharge and d_1 is the contracted depth of the jet. The measurements were taken in a 4 inch wide glass flume having an outlet weir for adjusting the downstream water level, the upstream depth being kept constant at 7.5 inches. The latter shows a number of photographs, exposure 0.2 sec, selected to describe the kind of patterns which occur

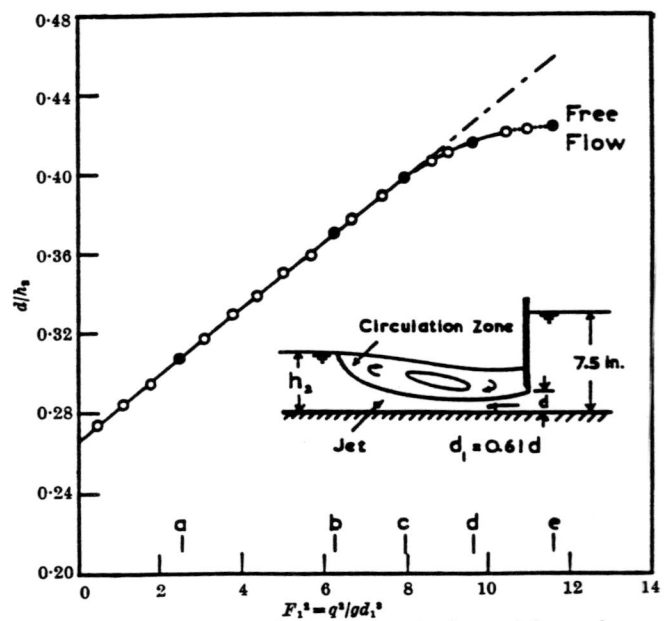

Fig. 1. Discharge characteristics of free and submerged flow under a sluice gate

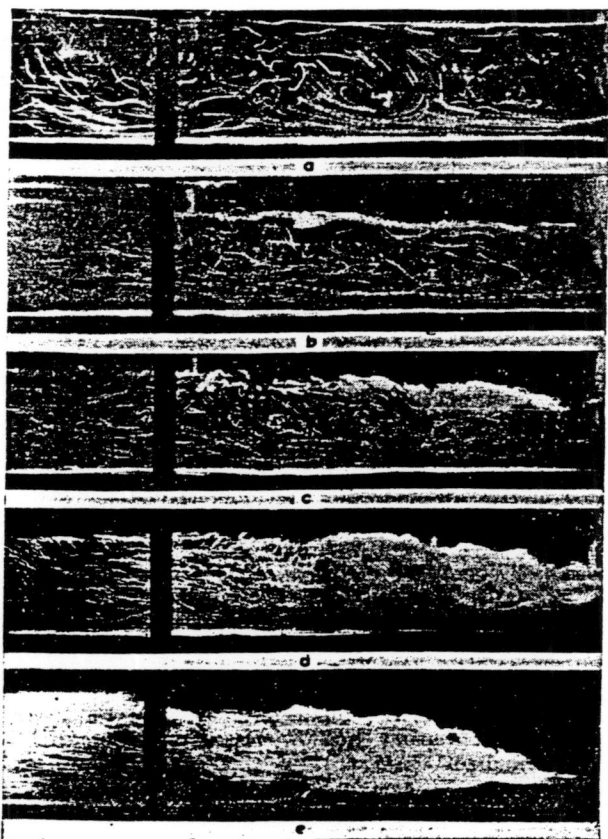

Fig. 2. Photographs of the flow patterns in the sluice-way jet for different discharges

under the particular conditions marked a-e in Fig.1. The photographs were taken with a 'Zeiss Ikon' camera of picture size 9 cm x 12 cm, with the flow illuminated by a mercury-vapour discharge lamp flashing at 100 times a sec. The tracers were droplets of a new white emulsion discovered experimentally by the author by mixing a mixture of olive oil and nitrobenzine, prepared to the same density of water, with a suitable percentage of water. Some idea of the scale of the photographs may be obtained by noticing the sluice opening which is 2 inches deep.

Discharge Characteristics

At a low rate of flow, the jet flows under the gate and apparently a sharp interface exists between the jet and the water above it (Fig.2a). This interface persists for a distance of several sluice openings but becomes contorted with billows and hollows as a result of the development of large-scale eddies between the jet and the overlying fluid. When the rate of flow is increased by lowering the downstream level, the jet boundary breaks up into large vortices at a point nearer the gate (Fig.2b). In this photograph, the diverging character of the jet is more

apparent and the length of the circulation is shorter. These are typical changes though the photographs are momentary. At the next higher rate of the flow (Fig.2c) the jet continues to behave in the same characteristic manner, though the large eddies now extend up to the free surface, with the result that a new surface effect comes into play: air is entrained at the occasional discontinuities produced by the strong vortex motions. Further, it is at this point that the straight line law of the discharge curve (Fig.1) terminates and transition begins and continues until the jet discharges freely into the air.

In the transition zone, a further increase in the rate of flow results in the formation of a train of vortices, as shown in Fig.2d. The vortices appear practically at the gate and they are packed with air bubbles, which sweel their volumes. With distance, the vortices appear to increase in bulk and to decrease in translational velocity, until at the end of the circulation zone they appear to lose distinction and are merged in the general turbulence. A region of violent turbulence thus extends throughout the flow the structural features of which, however, remain similar to those shown in Fig.2a, and resemble those of a wake flow behind a two-dimensional bluff body.

With decreasing downstream level, the depth at the gate becomes less and less until at a certain discharge the jet carries the water forward and the jet issues free from the gate. Diffusion then occurs further downstream under considerable pressure gradient, involving the phenomena called transitionally the 'hydraulic jump' (Fig.2e). The dotted part of the discharge curve, Fig.1, represents flows where the front of the jump is somewhere between the gate and the vena contracta section, about two gate openings downstream. It is interesting to note that the slope of the water surface, forming the upper boundary of the jump circulation zone, is about 1 in 4; this is practically the same as the rate of spreading of the outer boundary of a plane air jet.

Recently, I have investigated the diffusion of the jet under submerged conditions using a photographic technique. Owing to the entrainment of air bubbles, the flow under transition conditions could not be included. For the same reason, Rouse, Siao and Nagaratnam investigated the turbulence characteristics of the hydraulic jump by stimulating the flow pattern in an air duct.

References

1. Abramovich, G N. "The Theory of Turbulent Jets". MIT Press, 1963.
2. Rouse, H, Saio, T T and Nagaratnam, S J. J. Hyd. Div., Proc. ASCE, 1984, Part 1.

UNSTEADINESS OF CIRCULATION PATTERN

Introduction

When a turbulent jet flows through an enveloping fluid, the interface breaks up into vortices. The development of such vortices for the flow of a confined jet issuing from a sluice-way and diffusing into water downstream has previously been described[1]. The vortices grow as they flow downstream, and eventually extend throughout the entire height of the circulation zone. The mixing produced by these vortices causes the jet to expand until its upper limit momentarily reaches the free surface, where it divides and a stagnation point develops.

These vortices or large scale eddies occur randomly, and therefore some fluctuation in the position of the stagnation point is to be expected. Observation of the movement of fluid particles along the free surface indicated that these fluctuations were large and had far-reaching results; often they coincided with changes in form of the flow pattern, Fig.1. Pattern (c) is like an ordinary hydraulic jump, and also shows the unsteadiness characteristic of hydraulic jumps. At first, the change in form of flow was thought to occur only in semi-submerged patterns, but careful observations showed that the phenomenon is present, though it differs in magnitude, in all patterns. It was therefore very important to establish whether or not the fluctuation is the source of disturbance to the flow.

Experimental Method

Jets of dye were injected into the flow through two tubes, formed of two L-shaped lengths of hypodermic tubing placed back to back with injection openings 2 inches apart. The instrument was mounted on a carriage so that the dye jets were situated about 0.5 inches below the water surface of the flow pattern under examination. It was moved up and down the stream to search for the position of the momentary stagnation point, where a local hump formed on the free surface. Continuous observations over a period

of about 5 minutes (Fig.2) showed that the fluctuations were random, in the range from 0 to a maximum ± 15 per cent of the distance of the mean stagnation point from the start of flow (that is, ± 15 per cent of the length of the

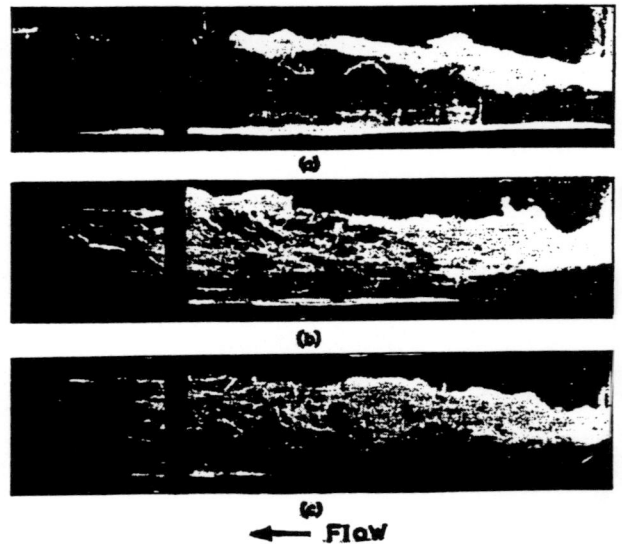

Fig. 1. Fluctuating forms of flow in the transition state of the sluice-way jet. Gate opening 2 in. Exposures 0·2 sec.

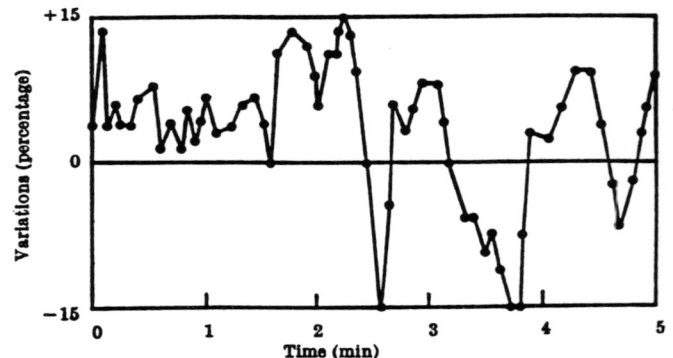

Fig. 2. Variations in length of circulation zone with time.

circulation zone), obtained by averaging the distance of the maximum and minimum positions. This range of fluctuations is confirmed by photographs of the flow patterns, a set of which is shown in Fig.3. The photographs are for different times under the same experimental conditions. The minimum and maximum positions of the stagnation point from the gate are about 22 and 30 inches, respectively; the latter position is shown in photograph (a).

Oscillation of Stagnation Point

One consequence of the movement of the stagnation point is a variation in the size and shape of the circulation zone, and this in turn causes changes in the form of flow. The actual mechanism by which a change in form occurs is not clear, but a possible explanation can be given: at first sight, on the basis of continuity, the quantity of fluid carried along by the jet must be equal to the quantity recirculated. Indeed, this balance is apporximately maintained during those periods when the large eddies of the flow form a nearly steady pattern, and during which

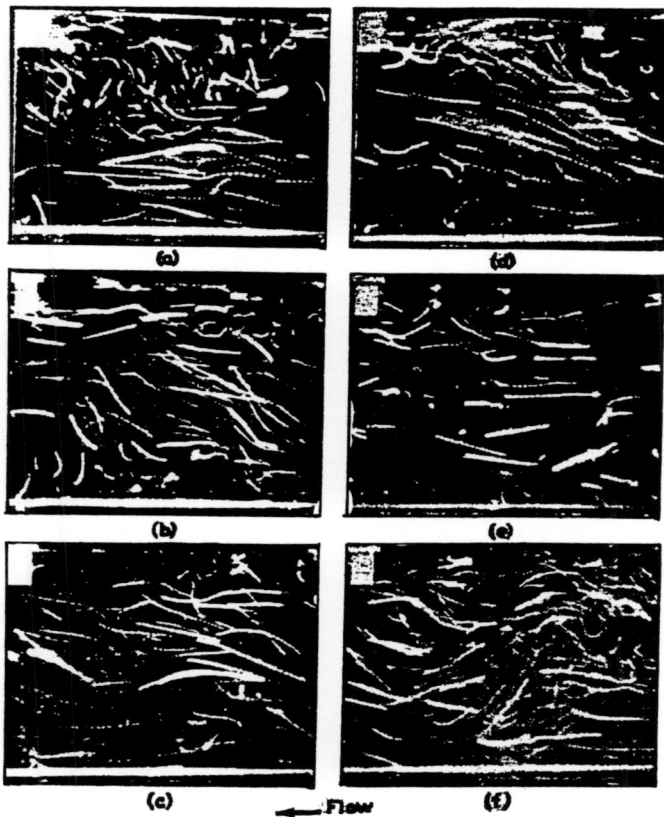

Figure 3: Flow near the nominal end of the circulation zone. Depth 4-8 in. Exposures 0.2 sec.

the length of the circulation l remains unchanged. This pattern is, however, often followed by another pattern in which the overlaying circulation is either shorter or longer than l.

When the shorter recirculation pattern forms, some of the water which would previously have been recirculated now passes downstream, causing simultaneously a decrease in the level of the water surface and a temporary increase in discharge. The lower water level also causes the velocity of the jet to increase, and if the mixing process remains unchanged, the amount of water entrained from the circulation zone increases. The shortening therefore leads to further reduction in the volume of the circulation, as shown by the deep trenches in Fig.2, and consequently the form of recirculation successively passes throught a sequence shown in Fig.1. The flow pattern in photograph (c) resembles that of the hydraulic jump, and since the overlying circulation of the jump is shorter than that in the submerged flow, this observation leads further support to the suggested interpretation.

Stabilization of Jet

Attempts were made to reduce the jet oscillation while retaining the general character of the flow. Essentially, this search was for simple ways of inducing permanent separation in such a way that it would deflect the diffusing jet away from the bed and towards the end of the circulation zone. Three methods were tried: (1) suction of the boundary layer, (2) application of an artifical bed roughness and (3) the use of sills as deflectors.

Experimentally, suction was applied in two ways. In the first, water was allowed to escape through five holes in the bed of the channel. These holes were of 0.2 inches internal diameter and were uniformly located across the width of a 1 ft. flume. To make use of these holes, experimental conditions were needed in which the circulation zones extended as far downstream as the holes. In the second technique, water was sucked away through a slit 0.5 inches deep made in a 1 inch diameter brass tube which was held on the bed with the slit facing upstream. Both techniques were investigated for a wide range of flow patterns, but the suction applied did not influence the flow in the boundary layer enougth to reduce the oscillation of the jet.

The second method used a coarse bed roughness to control separation by artificially varying the velocity distribution. In the presence of such a bed, it was expected that the point at which the thickness of the boundary layer is sufficient to become unstable and to cause the jet to separate from the bed and pass permanently over a bottom circulation zone might be reached sooner. Examination of the patterns in this flow, however, indicated no appreciable change in the state of the jet.

The principle of the third method is to create a permanent circulation on the bottom by lifting the jet away from the bed by means of a sill.

For this purpose, square and rectangular sills were first tried. These varied in height from 6 to 10 per cent of the downstream depth of the flow. Each sill was in turn held on the bed at a point towards the end of the circulation zone. Surprisingly, the presence of these sills had no influence whatsoever on the unsteadiness of the jet. A larger sill with a height of 20 per cent of the downstream depth was then tried. This sill created a zone of separation in the region immediately upstream, and at the separation point the jet was deflected away from the bed and passed permanently over the sill and then continued over a bottom circulation downstream of the sill. The oscillation of the jet was thus reduced, but the original pattern of flow was largely altered; because of this, this method was not pursued.

The failure of the smaller sills placed towards the end of the circulation zone to deflect the jet is understandable, because the boundary surface of the jet must have already been distorted by the convecton of the large eddies, particularly those in the second half of the circulation zone. The small sills would, therefore, prove effective only when located at a point where the flow is undistorted by large scale eddies with their associated pressure systems. Several points between the gate and the half-way point in the circulation were tried. In each case, the jet was deflected away from the bed, passed over a permanent bottom circulation, and created a noticeable hump in the water surface, except perhaps in the extreme case when the sill was placed at the half-way point in the circulation, and the hump was hardly detectable.

Reference

1.   Naib, S.K.A., Nature, 210, 694 (1966).

ABNORMAL GROUPING OF LARGE EDDIES

Introduction

It is well known that when a turbulent jet flows through an enveloping fluid, the surface of discontinuity breaks up into vortices which are carried alternatively along the surface. The development of these vortices in the course of flow of a jet issuing from a submerged sluice-way and diffusing into the water downstream has been previously described [1,2]. The vortices grow with distance and eventually extend to the water surface through the entire height of the circulation zone. The mixing produced by these vortices simultaneously causes the jet to expand until its upper limit momentarily reaches the free surface, where it divides and a stagnation point develops.

Development of Large Eddies

In the course of experiments to study the diffusion of the jet, it was observed that a particular kind of unsteadiness developed involving abnormal grouping of large eddies in the flow, a phenomenon which was clearly observed in semi-submerged patterns, as shown in Figure 1a. Here, two large eddies A and B are indicated near the middle of the circulation, where normally there would be a continuous return flow, Figure 1b. The photographs were taken with a "Zeiss Ikon" camera with the flow illuminated by a mercury-vapour discharge lamp flashing

Fig.1 - Two forms of flow downstream of a submerged sluice gate. Depth upstream of gate = 7.5 in. Downstream depth = 5.4 in. Exposures : 0.2 sec.

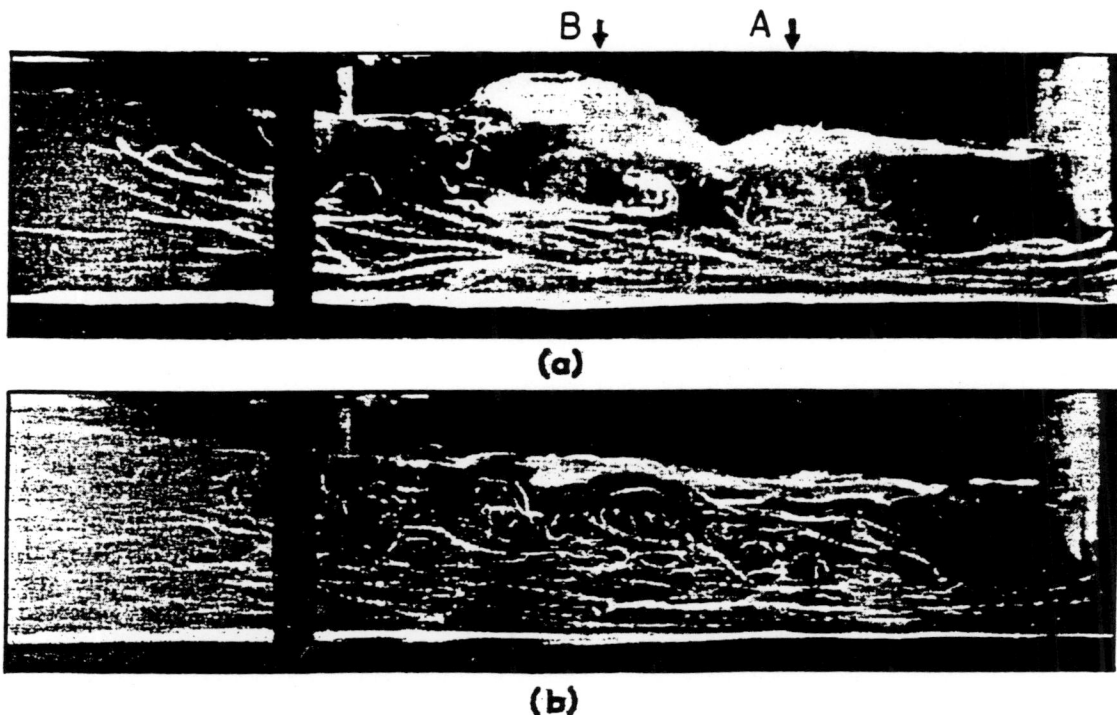

(a)

(b)

at 100 times a second [3]. The traces were droplets of a new white emulsion discovered by the author by mixing a mixture of olive oil and nitrobenzine prepared to the same density of water with a suitable percentage of water. Some idea of the scale of the photographs may be obtained by noticing the sluice opening which is 2 inches deep.

The situation exhibited in Figure 1 is very complex and we can only make the following guess about the physical sequence of events leading to it. Of the two eddies indicated in Figure 1a, the one nearer the gate appears to form first; it is probably caused by fluctuation either of the local turbulent stresses or of the pressure. Whatever the reason may be, once the eddy is initiated, momentarily a stream of fluid emerges from the jet into the circulation zone and proceeds towards A. This stream obstructs or retards the back flowing fluid and hence leads partly to an increase in the volume of fluid in Region B, and partly to the downward deflection and so ultimately leads to the formation of the second eddy B on the downstream side of A, thus creating two momentary circulation zones on top of the jet. The life either eddy is, however, short and is comparably with the
time it took to develop, and both eddies eventually pass downstream and disappear, after which the normal form of flow with one circulation zone overlying the jet, (Fig. 1b), prevails.

Fig. 2 - Plan of Ahlborn Tank

Ahlborn Tank Experiments

The above change in form of flow was also studied by photographing the motion of aluminium particles on the surface of water flowing through a sluice opening arranged in an Ahlborn tank [4] where a sharp-edged gate was arranged with one inch opening across two parallel walls placed plongitudinally in the tank in which water flowed from one end to antoher, Figure 2. The distance s between the two walls was varied from 2 to 8 times the gate opening d. To avoid the formation of surface waves due to surface tension, the flow adjusted to be less than 3 inches per second at the gate. The surface of water was illuminated by two 1000W lamps, and photographs were taken by a camera mounted vertically over the working portion of the tank. Figure 3, showing two of the resulting photographs for s = 8d, clearly confirms the abnormal grouping of large eddies and the formation of two momentary circulation zones on top of the jet.

Fig.3 - Two patterns of flow shown by the motion of aluminium flaken in an Alhborn tank. Width of flow = 8 in. Exposures : 1 sec.

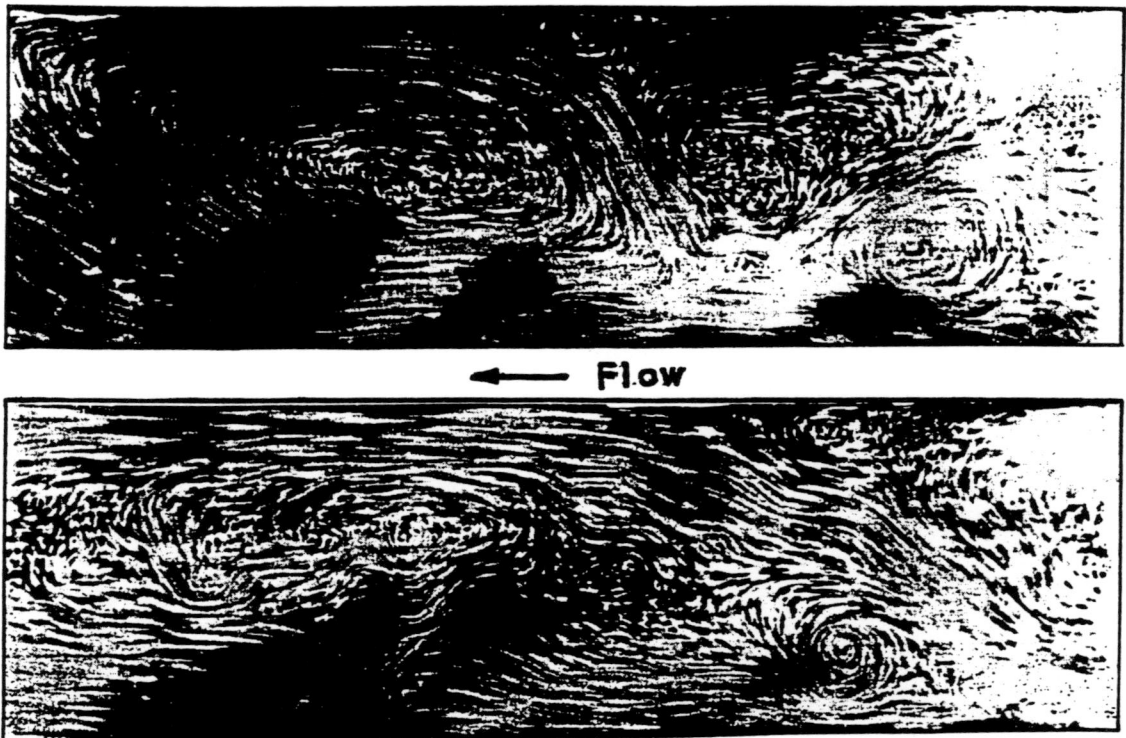

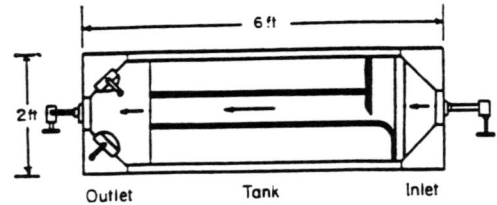

References

1,   Naib, S K A. "Flow Patterns in a Submerged
        Liquid Jet Diffusing Under Gravity". Nature,
        vol.210 (May 14, 1966), 694.
2.   Naib, S K A. "Unsteadiness of the Circulation
        Pattern in a Confined Liquid Jet". Nature, vol.212
        (November 12, 1966), 753.
3.   Naib, S K A. "Photographic Method for
        Measuring Velocity Profiles in a Liquid Jet". The
        Engineer, vol.221 (June 24, 1966), 961.
4.   Wallis, R P. "A Photographic Study of Fluid
        Flow Between Banks of Tubes". Proc. Inst.
        Mech. E., vol.142 (1939), 379.

PHOTOGRAPHIC METHOD FOR MEASURING VELOCITIES

Introduction

To measure photographically the magnitude and direction of turbulent velocities ranging from zero to 10 ft a second in a submerged water jet, where the turbulent components often momentarily exceed the mean velocity, is not an easy task. A survey of the literature disclosed the existence of a few techniques which could probably succeed and an investigation was undertaken to develop a method which would be most suitable for the specific problem.

The principle of any photographic method is that the motion of water at a point can be made visible by the presence of a suitable tracer which follows the stream. When this tracer is illuminated and photographed for a certain exposure time, it traces a track on the photograpic plate, showing by its length the instantaneous velocity, and by its orientation the instantaneous direction of flow at that point. Very few techniques for making such measurements have been developed during the last few decades. One of the earliest methods and indeed one of the most significant, is the method developed by Fage[1] in which he used an ultramicroscope for observing the motion of minute particles already present in the water. Another technique, developed by Kalinski,[2] used a motion picture camera for photographing oil bubles of the same specific gravity as water, which were illuminated by an intense beam of light. Both these techniques suffer from the disadvantage that they are developed for examination of a small region of the flow. Their modification to cover a large area of the flow is either improbable or at least would need a long period of development[3] and would require expensive equipments

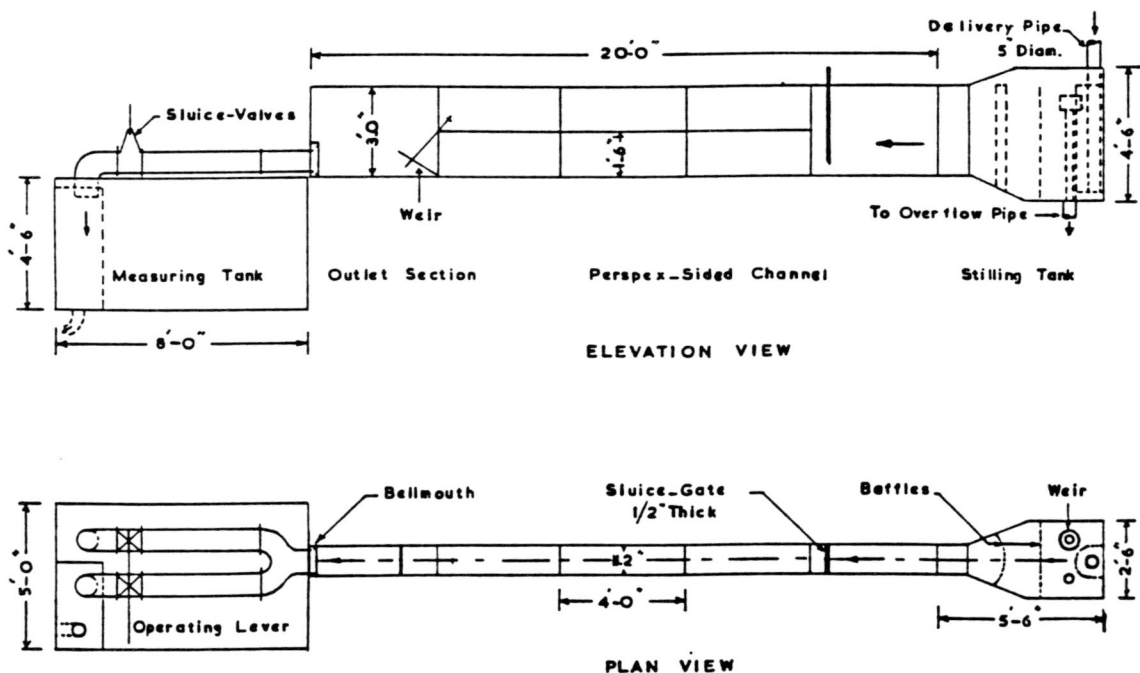

ELEVATION VIEW

PLAN VIEW

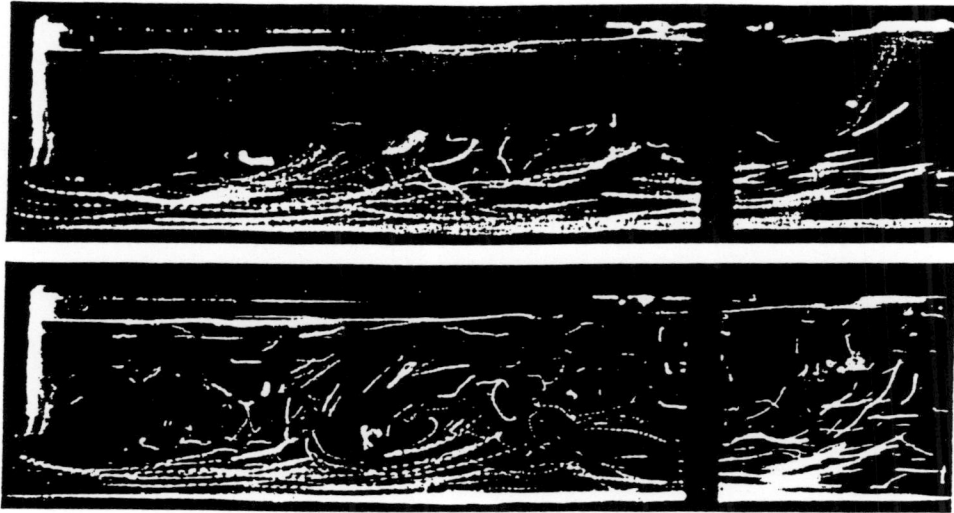

Fig. 1—Instantaneous photographs of flow downstream of a submerged sluice gate

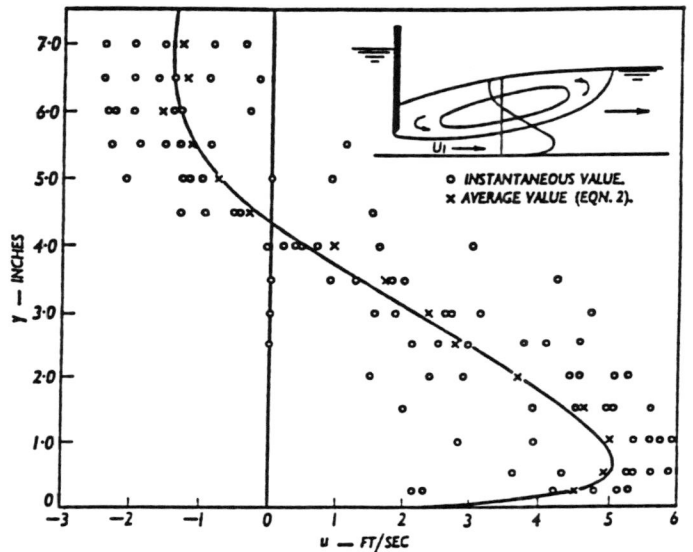

Fig.2 - Record of fluctuating velocities and the mean velocity profile

which could not be met with the facilities at the disposal of the author. Having thus established the need for a new photographic technique, there were two problems to be solved: one of developing a simple illumination system, and the other of finding a suitable tracing material.

Illumination

The illumination of a steady two-dimensional flow in a rectangular channel can be achieved by the use of a beam of continuous light projected through the middle of the flow. The lengths of the tracers passing through the beam of light, are then recorded correctly on the photographic plate. However, such photographs become increasingly obscure as the degree of turbulence and cross-wise motion increases. For these flows, the beam width must exceed the lateral distance travelled by the tracers during the exposure of the plate if the track lengths are to be determined accurately. On the other hand, an excessively wide beam covering most of the channel width, leads to a confused photograph.

One way of overcoming this difficulty is to use a timing shutter in conjuction with a continuous light source. In this technique, the tracks on the photographic plate are interrupted either by interposing a periodic obstruction between the reflected light from the tracers and the camera, or by chopping the light on its way to the tracers. This gives a series of dashes on the photographic plate, from whcih velocities can be deduced. With this in mind, the flow, with a black background, was illuminated by a

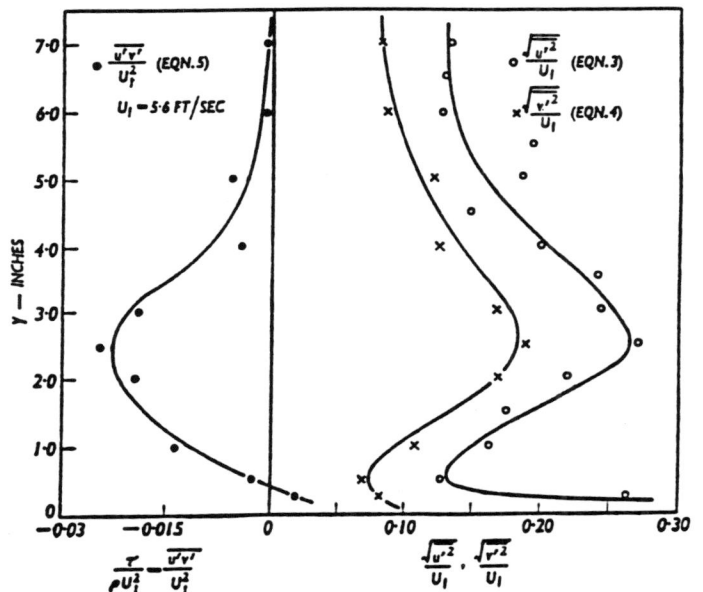

Fig.3 - Distirbution of turbulent intensities and shear stress for the velocity profile in Fig.2.

105

500W Mazda lamp, whose reflector with a slit arranged just above the water surface by two suitably bent plates, gave a clearly defined beam of light half an inch thick. The periodic obstruction between the tracers and the camera consisted of a disc 18 inches diameter, and with twelve equally spaces slots each a quarter of an inch wide by 6 inches deep. This disc was mounted on the spindle of an electric motor whose speed was controlled by a rheostat.

Though practicable, this method proved cumbersome to use and it was discontinued.

Another method of overcoming the difficulty is to use a mercury-vapour discharge lamp. This has the advantage that it provides virtually continuous illumination but with a pulse from the alternating main supply, which provides a time base. The lamp used was 120W Philips discharge lamp with its bulb removed. The light element was located 4 inches below a 10 inch diameter glass reflector to produce a sheet of light half an inch thick by 8 inches long as described above. This method proved to be satisfactory.

Tracing Material

Solid tracers which have been used in liquids include aluminium particles, polystyrene particles, wax spheres with density raised by copper fillers, and sawdust. Each of these, however, suffer from certain disadvantages. The first two introduce gravitational errors in tracing the paths of the flow; while the latter two are not easily observed by reflected light. Liquid tracers include mixtures of olive oil and nitrobenzene, and white spirit carbon tetrachloride suitably coloured by waxlyne dye. Such mixtures were prepared to have the same density as water and were introduced by means of a syringe into the inlet tank. With these droplets, well defined tracks could only be seen in the regions of the flow where the velocities were less than 1.5ft/sec. It was found that the migration of minute organisms, air bubbles and the large number of droplets produce conditions which detract considerably from the degree of constant between the yellowish oil bubbles and the surrounding water. Even when the condition of the water was improved and a more powerful illumination was used, the photographic measurements could only proceed in a very slow and tedious manner. This situation was changed when the author discovered experimentally that when a mixture of olive oil and nitrobenzine, prepared to the same density of water, is mixed with a suitable percentage of water, it produced a new white emulsion whose bubbles proved ideal for photographic measurements. When these bubles were illuminated, a fairly high proportion of the light emerged normal to the incident beam, and the particles were seen very clearly in this direction. This tracer has the additional advantage over other known tracers of satisfying exactly the denstiy requirement. With a Zeiss Ikon camera of picture size 9cm x 12cm and using Kodak P 1200 plates, excellent photographs of the flow were obtained, from which instantaneous velocity profiles were measured. Fig.1 shows photographs of the fluctuating patterns in a jet issuing from a submerged sluice-opening, 2 inches deep, and diffusing into the water downstream under constant discharge. The exposure time was one-fifth of a second, and the flow far downstream of the gate was illuminated by a continuous light.

Measurements From Photographs

At any point in the flow, the turbulent motion may be decomposed into mean flow with steady horizontal and vertical velocity components u, v and a turbulent motion with velocity components u^1 and v^1. The velocity components of the total motion will then be

$$U = u + u^1 \text{ and } V = v + v^1$$

By using stroboscopic light, the paths of the illuminated oil bubbles over successive periods of one-hundredth of a second were obtained as a series of dashed streaks, and by comparing the length between the centres of adjacent dashes with a linear scale included in the photograph, the magnitude of the instantaneous vector and hence the components of velocity U and V parallel to the photographic plane were obtained.

For obtaining average values from the instantaneous values, three methods are available:

(1)   When the motion at a particular point in the flow is recorded cinematographically over a period of time then the mean velocity is given by

$$u = \lim \frac{1}{T} \int_o^T U dt$$

where T is the total period of the record.

(2)   When successive values of the velocity at the point are recorded, then with S as the order of measurement and N the total number of the intermittent measurements, the mean value of the velocity is

$$u = \lim \frac{1}{N} \sum U_s$$

(3)   The mean value is taken as the arithmetic average of the minimum and maximum observed values of the instantaneous component velocity.

In the present investigation, only the last two methods could be used, and these often gave nearly equal values; however, the final velocity profile was drawn for best representation of the plotted points as shown in the sample curve, Fig.2. Here, the measurements refer to a section 1.5ft downstream of a sluice-way, 3 inches deep, passing a discharge of 0.77 cusecs in a 1ft wide flume, the downstream depth being 9 inches.

Once the mean values of velocities have been determined, some information about the characteristics of turbulent motion can be obtained. The difference between each instantaneous value, and the mean value gives the fluctuation, u^1. The root mean-square value of the turbulent fluctuation is then

$$\sqrt{u^{1^2}} = \left[\lim \frac{1}{N} \sum u_s^s\right]^{\frac{1}{2}}$$

Similarily, we can write

$$\sqrt{v^{1^2}} = \left[\lim \frac{1}{N} \sum v_s^2\right]^{\frac{1}{2}}$$

From the fluctuations u and v, the turbulent shear stress

$$\tau = -\rho u' v'$$

may also be calculated. Since

$$\frac{1}{4}\left[(u'-v')^2 - (u'+v')^2\right] = u'v'$$

then the mean value of the product $-u$ v, is

$$-u'v' = \lim \frac{1}{N} \sum (-u'v')_s$$

where ρ is the density of the fluid. For a reliable analysis of velocities, a large number of photographs, up to twelve and with a total of about 150 readings, were taken to include all the different instantaneous velocity profiles of the fluctuating flow at each section. Typical results are shown in Fig.3. It is interesting to note that measurements by Rouse and co-workers[4] in an air model of the hydraulic jump using the hot-wire technique show similar distributions of turbulence characteristics.

References

1. Fage, A. "Studies of Boundary Layer Flow with a Fluid Motion Microscope". 50 Jahre Grenzchilichtforschung, Editors H. Gortler and W. Tollmien, Page 132, 1955.
2. Kalinske, A A. "Relation of the Statistical Theory of Turbulence to Hydraulics". Trans. ASCE, Vol 105, Page 1547, 1940.
3. Winter, E F and Deterding, J H. "Apparatus and Techniques for the Application of a Water Flow System to the Study of Aerodynamic Systems". British Journ. of Applied Physics, Vol 7, No.7,p247, July 1956.
4. Rouse, H. Siao, T T and Nagaratnam, S. "Turbulence Characteristics of the Hydraulic Jump". Journ. Hyd. Div. Proc. ASCE, Vol 84, Part I, p1528, 1958.

DIFFUSION OF THE SLUICE WAY JET

Introduction

When a stream passes under a submerged sluice gate into water on the downstream side, it forms a jet which goes far downstream before it is dispersed [1]. A knowledge of the behaviour of such a jet is of engineering importance because it allows designers to predict the rate of deceleration of the jet which is necessary in the prevention of scour downstream of hydraulic structures. Furthermore, it leads to a better understanding of the mechanism of turbulent diffusion of jets in restricted spaces. Although it was one of the classical problems in hydraulics it has been little studied in the past. Early publications mainly describe model tests for guiding design and predicting the depths and heads associated with the discharge characteristics. Consequently the correlations were given in terms of one-dimensional forms of continuity, momentum and energy relationships [2].

The approach followed by these investigators has been, in general, empirical in character concerning an overall study of the flow variables. While information of this type is essential for industrial design, it does not lead far towards a fundamental understanding of the flow mechanism. However, when as in the present investigation the details of flow are examined a definite picture emerges of the jet dispersing in the deeper water downstream, in the manner occurring in other diffusion problems and conforming with similar statistical picture.

Only a little relevant work has yet been published. Liu [3] investigated by pitot-cylinder the mean patterns with different depths of water downstream. He found that the distance required for the jet to diffuse in a given depth is independent of the Froude number. By using a pitot-tube operating an electrical circuit, Henry [3] measured the longitudinal component of turbulence in certain of the patterns previously studied by Liu. He discovered that the energy of the turbulence is less than one per cent of the energy of the mean flow.

Their results contained discrepancies such as the one between the measured discharge and that calculated through the integration of the velocity profiles. Rajaratnam [4] attempted to analyze the jet as the case of a plane turbulent wall jet under an adverse pressure gradient over which a backward flow has been placed. He presented the forward flow simply as a plane wall jet and evaluated the backward flow using the results of Liu and Henry. The two parts were joined to predict the characteristics of the jet. Unfortunately, the theoretical profiles did not compare at all with the measured surface profiles, either in shape or magnitude [5]. The above three works, while containing some unexplained discrepancies, nevertheless clearly contribute to a better understanding of the general flow pattern.

As a further contribution, the present investigation was undertaken with a view to overcoming the difficulties of the above investigators. To this end a preliminary series of instrumentation tests were undertaken, followed by extensive measurements to study the flow which proved very complicated and unsteady [6,7], requiring the development of a photographic technique to measure velocities [8]. When this was achieved, distributions of mean velocity were determined in various sections of the jet for differing ratios of the limiting downstream depth to sluice opening for the values 2,3,4, 5.5 and 8 at a fixed Froude number of the flow.

Experimental Apparatus and Method

The experiments were carried out in a perspex flume, 300mm wide by 900mm deep and 6m long, fitted with a sharp edged sluice gate 12.5mm thick. To damp the surging and turbulence, the water entered the inlet reservoir through a central perforated pipe which was shielded by a curved sheet and additional baffles were inserted across the tank. Irregularities in the discharge were reduced by a cylindrical overflow weir. An adjustable weir near the end of the flume regulated the downstream level. The discharge was measured in a calibrated tank. Gate openings, d, of 25mm, 50mm and 75mm were used. The Froude number of the flow $F = (qd_o^3)^{0.5}$ where q is the unit discharge and $d_o = 0.61d$, was kept constant at 2.3 for all the flow patterns. Because of an inherent fluctuation of the flow pattern, the velocity distributions in the jet were measured by a photographic technique developed by the author [8]. A mercury-vapour discharge lamp was used to produce a sheet of light 10mm thick by 200mm long along the centreline of the flume. By using a stroboscopic light, the paths of the illuminated oil tracers over successive periods of one-hundredth of a second were obtained as a series of dashed streaks. By comparing the length between the centres of adjacent dashes with a linear scale included in the photograph, the magnitude of the instantaneous vector and hence the horizontal and vertical components of velocity parallel to the photographic plane were obtained. When successive values of the velocity at a point were recorded, then with S as the number of measurement and N the total number of intermittents measurements, the mean value of the velocity was found by

$$u - \lim \frac{1}{N} \sum U_s$$

The final velocity profile was then drawn for the best representation of the plotted points.

Description of Flow

Although the physical aspects of the dispersion of a sluice jet have some similarity with the simpler problem of the dispersion of a jet issuing into infinite fluid, it has several fundamental properties which distinguish it. Apart from the contraction due to the sharp edge of the gate, the differences are those described below.

The simple concept of a jet issuing from a point source into an infinite surrounding is no longer appropriate on account of the much larger size of the sluice opening in comparison with the dimensions of flow; the width of opening may be half of the downstream depth. A schematic section of the flow is shown in Fig.1. The jet issues into a slow moving medium and is guided by a

lower boundary. The mixing process between the jet and the surrounding fluid causes part of the latter to be carried forward with the jet under conditions in which not only the forward momentum but also the total discharge is conserved. The process of entrainment leads to recirculation to replace the fluid entrained. Expressed in another way, the action of the jet is to produce a low pressure region around it which is then fed from regions of higher pressures downstream. Thus a zone of circulatory motion is established which is usually termed a "roller". Meanwhile, the mixing process causes the jet to expand vertically until the upper mean flow line reaches the free surface. This point is the stagnation point, which defines the downstream end of the roller.

In comparison with a free jet, the chief difference is that in the sluice jet the upper stream or limit of the flow is a free surface which slopes upward in the downstream direction. Since this surface is nearly free from shear stress, its slope is evidence of positive pressure rise in the conversion of the high velocity jet into pressure. The amount of the resulting pressure difference depends on the sluice openings, the discharge and the downstream depth of the flow, as indicated in Figs.2 and 3. The former concerns variation in downstream depth, the upstream head being constant; whilst the latter shows a number of photographs selected to describe the kinds of patterns which occur under the particular conditions marked, a, b, c, d and e in Fig.2.

Water Surface Profiles

Figure 4 shows the mean depth of water plotted non-dimensionally against distance for five patterns of flow. It is seen that the depth h at first decreases from its value at the gate to a minimum value h_o at some distance downstream. This is partly due to the stagnation at the gate together with effects due to contraction of the jet. Further downstream, the depth gradually increases until it reaches the final downstream level. For the semi-submerged patterns, the rate of increase in depth is seen to diminish with distance downstream and since the shear stress at the surface is negligible its slope indicates considerable conversion of kinetic head into potential head

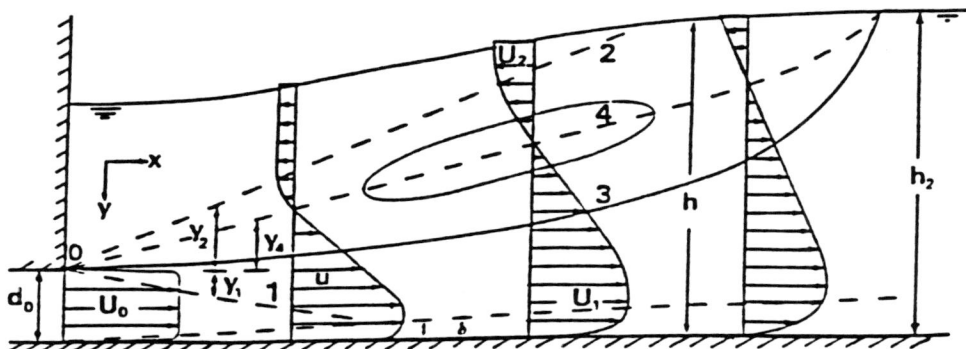

Fig. 1

and this is seen to occur mostly in the first half of the circulation zone.

In the analysis of the half-jet boundary, Tollmien [9] found that along the streamline $\psi=0$, forming the upper boundary of the jet, the velocity is 0.566 U_o. Since this streamline terminates at the stagnation point at the downstream end of the circulation zone, the velocity head $0.32\ U_o^2/2g$ must be partly converted into pressure head which is $(h_2 - h_o)$, and the rest is dissipated in turbulence. The proportion of pressure recovery varies with the degree of submergence of the jet as indicated by the comparison of the calculated and measured heads given in Table 1 below:

Table 1 - Recovery of Pressure Head

h_2/d	2.0	3.0	4.0	5.5	8.0
$0.32U_o^2/2gd_o$	0.293	0.262	0.159	0.111	0.056
$(h_2 - h_o)/d_o$	0.228	0.259	0.129	0.070	0.030

In the semi-submerged patterns it is surprising to find that at the boundary of the jet with the overlying circulation, where the shear stresses are maximum, there is relatively little head lost in turbulence and most of the velocity head is converted into pressure head. For the highly submerged patterns Tollmien's theory is less applicable as the length of the potential core is small compared to the length of the circulation zone.

Spreading of Jet Boundaries

The curves of Figs.5 and 6, show the spreading of the jet boundaries for various ratios of submergence. The angles of spread are compared in Table 2 with those of other confined jets with recirculation zones and also with the classical results of Tollmien for the plane free jet. The following conclusions may be derived:

(i)   The rate of growth of the boundary layer δ near the bed of the channel is almost linear with an inclination of 0.5° in the entrainment region. Over most parts of the wall layer, the Reynolds number of the flow is greater than 5×10^5, indicating that the flow is turbulent.

(ii) The inner boundary 01 of the potential core expands at an angle of 4.8° or about 1 in 12, which is the same as that for a half-jet boundary.

(iii) The mean spreading angle of the outer boundary 02 is 15.3° which closely conforms to Tollmien's solution for a plane free air jet with an expansion angle of 14° or about 1 in 4. The section where 02 reaches the water surface marks the end of the entrainment region and the position of maximum reverse velocity in the circulation zone above the jet.

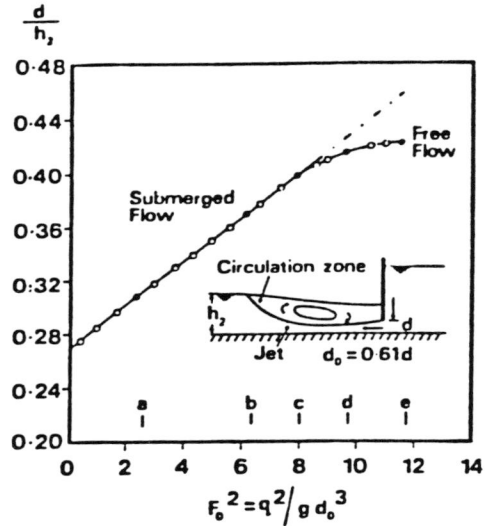

Fig. 2

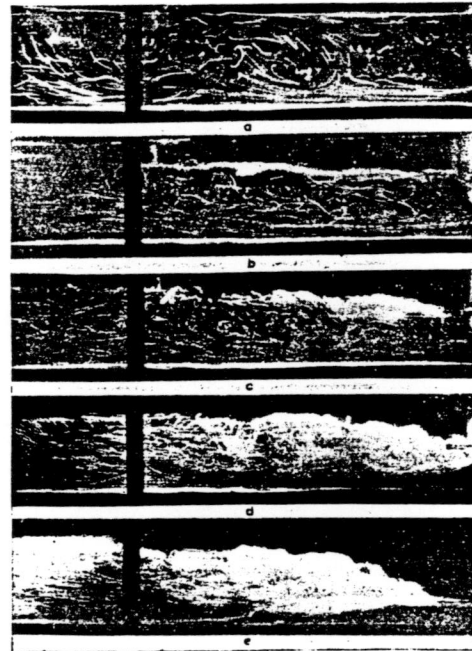

Fig. 3

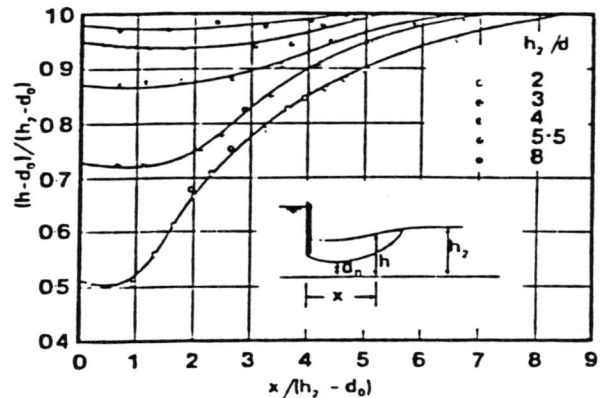

Fig. 4

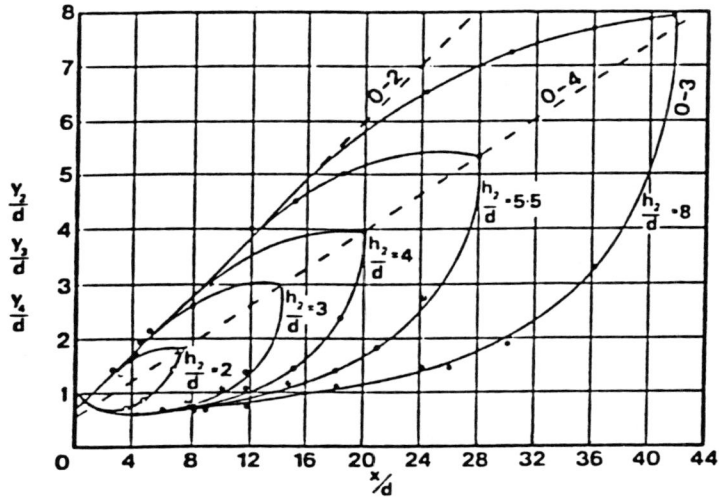

Fig. 5

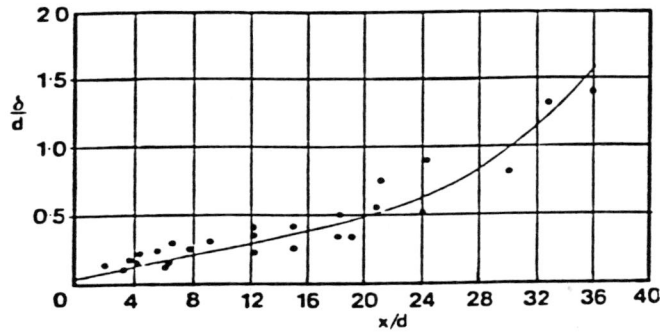

Fig. 6

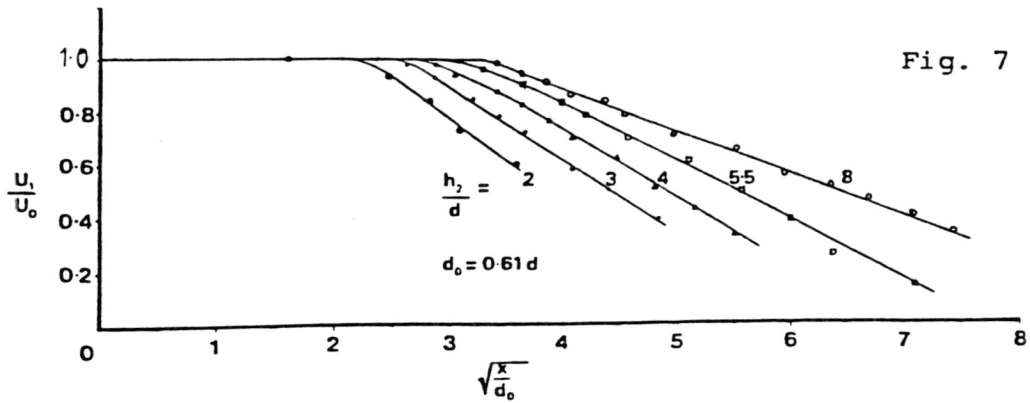

Fig. 7

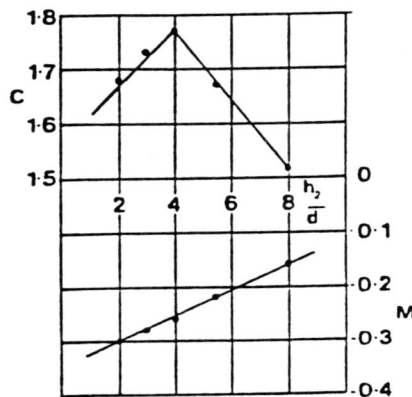

Fig. 8

Table 2 - Spreading of Jets with Circulation Zones

FLOW	BOUNDARY			
	01	02	03	04
Diffusion of Sluiceway Jet Naib (Present Data)	4.8°	15.3°	1.3°	9.8°
Surface Motion of Plane Liquid Jet Naib [10]	6.4°	15.3°	1.3°	9.4°
Rectangular Channel Expansion Naib [11]	5°	13°	1.2°	7.5°
Confined Wall Jet Forthmann [12]	5°	15.2°	1.5°	8.9°
Flow Behind Bluff Body Abramovich [13]	6.3°	14°	1.3°	9.4°
Half Jet Boundary Tollmien [9]	5°	-	0.91°	9.8°
Plane Free Jet Tollmien [9]	5°	15°	-	-

(iv) The boundary of the jet 03 defined by $\psi=0$ increases gradually at an angle of 1.3° until near the end of the circulation zone it curves sharply upward to reach the free surface, whereby the stagnation point is reached.

(v) As the submergence increases, the distance of the stagnation point from the gate increases, the locus being approximately a straight line with a slope of the boundary 04 being 9.8° or about 1 in 5.83 which is practically the same as predicted by Tollmien for a half jet boundary. Additional experiments with different Froude numbers of the flow gave the same results indicating that the length of the circulation zone is only dependent on the degree of submergence.

Decay of Maximum Velocity

The longitudinal distributions of the maximum velocity U_1 divided by the maximum jet velocity U_o are plotted against $(x/d_o)^{0.5}$ in Fig.7. The curves, each of which represents the results of one pattern, show that the maximum velocity remains constant for a certain region downstream of the gate. The length of this potential core region, x_c, as given by the horizontal part of each curve, is variable depending on the submergence as shown in Table 3.

Table 3 - Length of Potential Core

h_2/d	2	3	4	5.5	8
x_c/d_o	3.3	4.4	5.6	6.5	8.2

Following the core region is a transitional length, which is characterised by a gradual decrease of the maximum velocity. This zone opens into the fully developed turbulent region where surprisingly the jet decays linearly with the square root of the distance x, as compared to an inverse variation for the plane free jet. The straight lines indicate that for both the entrained and recirculated regions

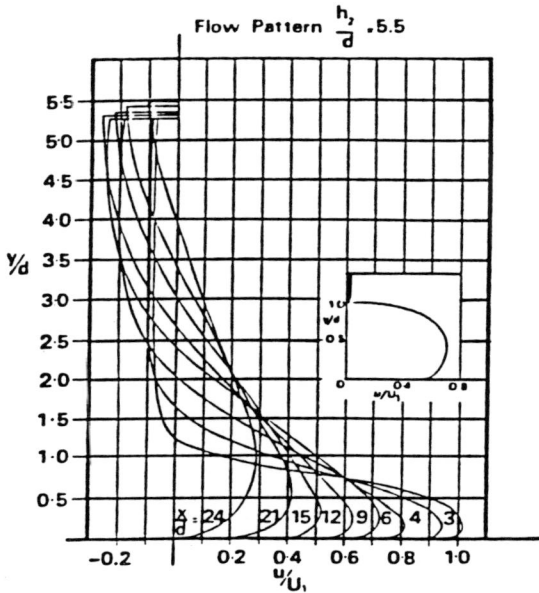

Fig.9

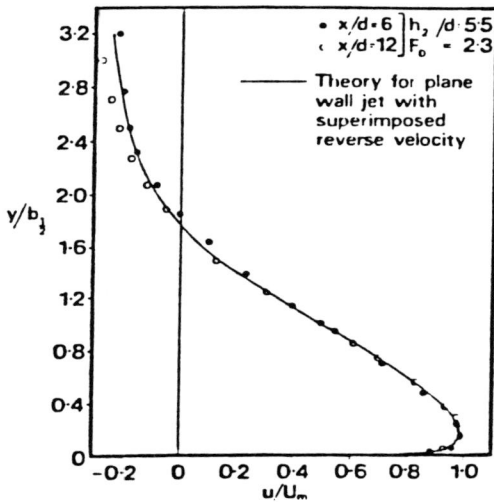

Fig. 10

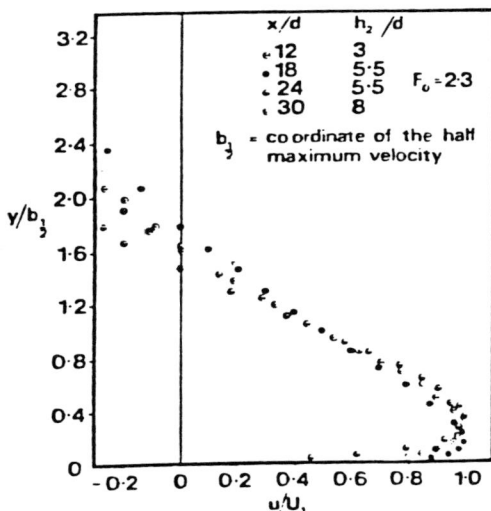

Fig. 11

of the jet, the decay of maximum velocity is governed by the same power law, as given by the following general equation:

$$U_1/U_0 = C - m \sqrt{x/d_0}$$

The variations of c and m with h_2/d are given in Fig.8, where it is seen that the value of C increases to a maximum of 1.77 for $h_2/d = 4$ and then decreases to 1.52 for $h_2/d = 8$. The straight line corresponding to $h_2/d = 8$ lies highest and the rate of the decay is least, indicating that mixing takes place most slowly in the highly submerged case. For the present experimental conditions the following equations have been derived:

$$h_2/d = 2.0 \; ; \; U_1/U_0 = 1.68 - 0.30 \sqrt{x/d_0}$$
$$h_2/d = 3.0 \; ; \; U_1/U_0 = 1.73 - 0.28 \sqrt{x/d_0}$$
$$h_2/d = 4.0 \; ; \; U_1/U_0 = 1.77 - 0.26 \sqrt{x/d_0}$$
$$h_2/d = 5.5 \; ; \; U_1/U_0 = 1.67 - 0.22 \sqrt{x/d_0}$$
$$h_2/d = 8.0 \; ; \; U_1/U_0 = 1.52 - 0.16 \sqrt{x/d_0}$$

Mean Velocity Profiles

Distributions of mean velocity for one pattern $h_2/d = 5.5$, non-dimensionlised by the maximum jet velocity, are plotted in Fig.9. The variation of the velocity distribution with respect to the distance downstream was found to be substantially the same for all the flow patterns and is described typically as follows.

At the gate, the distribution is characterised by a constant velocity up to the middle of the sluice opening, decreasing above this, apparently in a parabolic manner, to zero at the tip of the gate. Downstream of the gate, the jet undergoes a contraction, while entraining the surrounding fluid. As a result the velocity distribution near the vena contracta section is marked by an increased maximum velocity which remains constant up to a depth of 0.4d. Further downstream, the flow is fully developed. In the entrainment region, Fig.10, the profiles lie very nearly on one curve and therefore are similar despite the different rates of pressure rise under various degrees of submergence. The profiles are also seen to be in good agreement with the theoretical profile for a plane wall jet [14] with a superimposed reverse velocity of $U_2 = 0.25 \, U_1$. In the recirculation region, Fig.11, the profiles vary from section to section and therefore are no longer similar. This lack of similarity is due to different characteristics of the flow in this region, such as the increasing curvature of the streamlines as the stagnation point is reached and the excess turbulence generated due to large time variations in the length of the circulation zone up to $\pm$ 15 per cent. Photographs of the flow in this region show a distinctive pattern of large eddies which because of their size, convective motion and random formation cause the jet to oscillate both vertically and horizontally. These changes are often accompanied by separation of the boundary layer along the bed of the channel. It appears that such a large scale motion carries a mixture of smaller eddies distributed across the flow and so intensifies the mixing. This in turn causes the jet to decay much faster than in the entrainment region.

References

1. Naib, S K A: Flow patterns in a submerged liquid jet diffusing under gravity. Nature, Vol 210, p694, May 14th 1966.
2. Addison, A: Hydraulic Measurements. Chapman Hall Ltd, 1948.
3. Henry, H R: Discussion on diffusion of submerged jets. Proc. ASCE, Vol 75, p1541, 1949.
4. Rajaratnam, N: Submerged hydraulic jump. J.Hyd.Div., ASCE, Vol 91, July 1965.
5. Ramaprasad and Ramanoorthy, M V: Discussions on submerged hydraulic jump. J.Hyd.Div., ASCE, Vol 92, Feb 1966.
6. Naib, S K A: Unsteadiness of the circulation pattern in a confined liquid jet. Nature, Vol 212, p753, November 12th 1966.
7. Naib, S K A: Abnormal grouping of large eddies in a submerged liquid jet. La Houille Blanche, No 3, p282, 1967.
8. Naib, S K A: Photographic method for measuring velocity profiles in a liquid jet. The Engineer, Vol 221, p961, June 24th 1966.
9. Tollmien, W: Calculation of turbulent expansion processes. NACA T M, No 1085, 1945.
10. Naib, S K A: Surface motion of a plane liquid jet. La Houille Blanche, No 6, p377, 1980.
11. Naib, S K A: Mixing of a subcritical stream in a rectangular channel expansion. J I Wat. E, p199, May 1966.
12. Forthman, E: Uber Turbulente Strahlausbreitung. Ing.Arch. Bd, p42, 1934.
13. Abramovich, G K: The theory of turbulent jets. MIT Press, Boston, USA, 1963.
14. Glauert, M B: The wall jet. J.Fluid Mechanics, Vol 1, p625, 1956.

PART VIII

JET DISSIPATION
STRUCTURES

Hydraulic Energy Dissipators
Talus Protection Works

HYDRAULIC ENERGY DISSIPATORS

Introduction

When water flows over a spillway or a weir, it leaves the sloping face as a jet with a velocity which is much higher than that of the natural stream. If the jet is unchecked, it will erode the unprotected bed of the downstream channel, thus endangering the safety of the structure. An arrangement must, therefore, be designed where the energy of the jet is dissipated and where the flow changes from the supercritical to the subcritical state before it enters the downstream channel. Such a transition is known as an energy dissipator or a stilling basin. Since the change in flow condition takes place in a hydraulic jump, the basic dimensions of hte basin are governed by the characteristics of this phenomena.

The formation of the jump at the toe of the spillway requires that the tailwater depth in the downstream channel be equal to the conjugate depth of the jump. If the tailwater depth is greater, the jump will move up the sloping glacis and will be drowned; however, this condition is hydraulically safe. If, on the other hand, the tailwater depth is smaller, the jump will move downstream to a location where channel friction has decreased the velocity of the jet and increased its depth to a value consistent with a jump having the tailwater depth. This is a dangerous condition and its remedy lies in providing sufficient depth to create a jump near the toe of the spillway. The required depth can be obtained either by the construction of a small weir below the apron, or by depressing the elevation of the stilling basin. Except under very low tailwater levels, the latter method is widely used. The theory and design of this type of stilling basin is given in the next section.

Theory and Design

Consider the flow in a channel drop structure shown

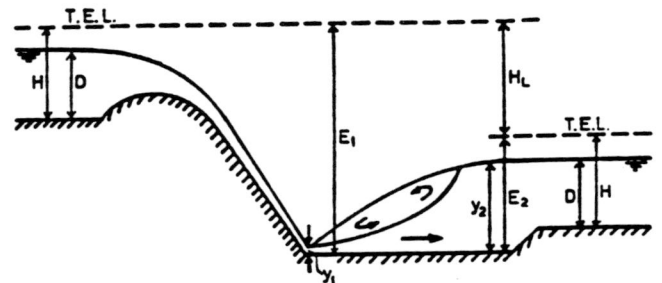

Fig. 1—Flow in a channel drop structure

in Fig.1. In practice, the upstream and downstream bed levels, the depth of flow D, the head loss H_L and the unit discharge q are known. It is required to calculate y_1, y_2 and the elevation of the stilling basin. According to Newton's second law of motion, the rate of change of momentum of the stream must be equal to the difference between the hydrostatic forces corresponding to the depths y_2 and y_1.

$$\rho q \left(v_1 - v_2\right) = \frac{1}{2}\, \rho g \left(y_2^2 - y_1^2\right) \qquad (1)$$

where ρ is the density of the fluid. Substitution of the continuity relationship, $q = y_1\, v_1 = y_2\, v_2$, yields the expression

$$y_1 y_2 \left(y_1 + y_2\right) = \frac{2}{g}\, q^2 \qquad (2)$$

The solutions of this equation for y_1 and y_2 are:

$$y_1 = -\frac{y_2}{2} + \sqrt{\frac{2q^2}{gy_2} + \frac{y_2^2}{4}} \qquad (3)$$

116

and

$$y_2 = -\frac{y_1}{2} + \sqrt{\frac{2q^2}{gy_1} + \frac{y_1^2}{4}} \qquad (4)$$

Substitution for the Froude Number $F_1 = q/(gy_1)^{0.5}$ in equation (2) and rearranging, we get

Fig. 2 — Curves to determine E_2

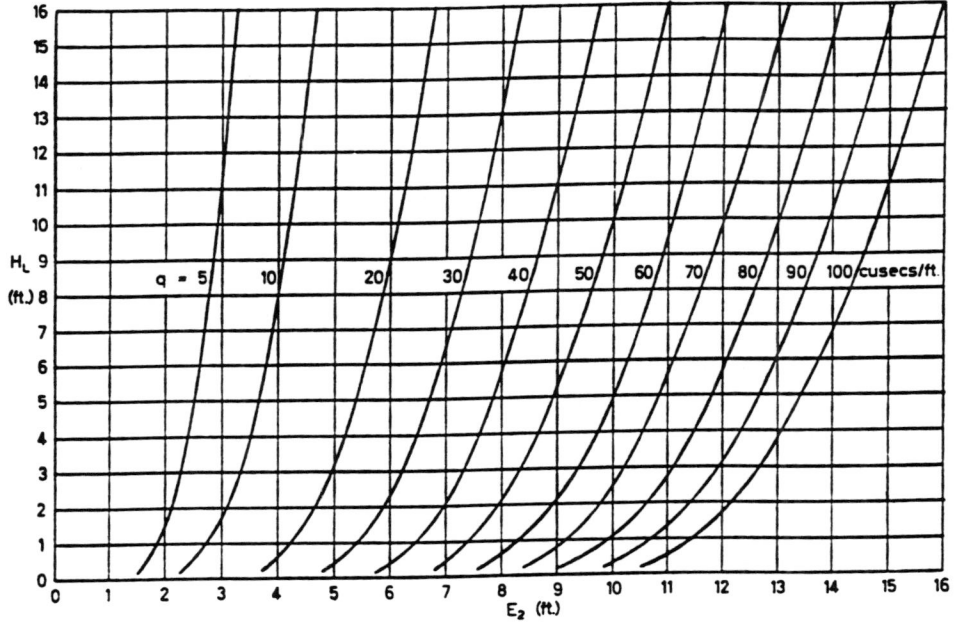

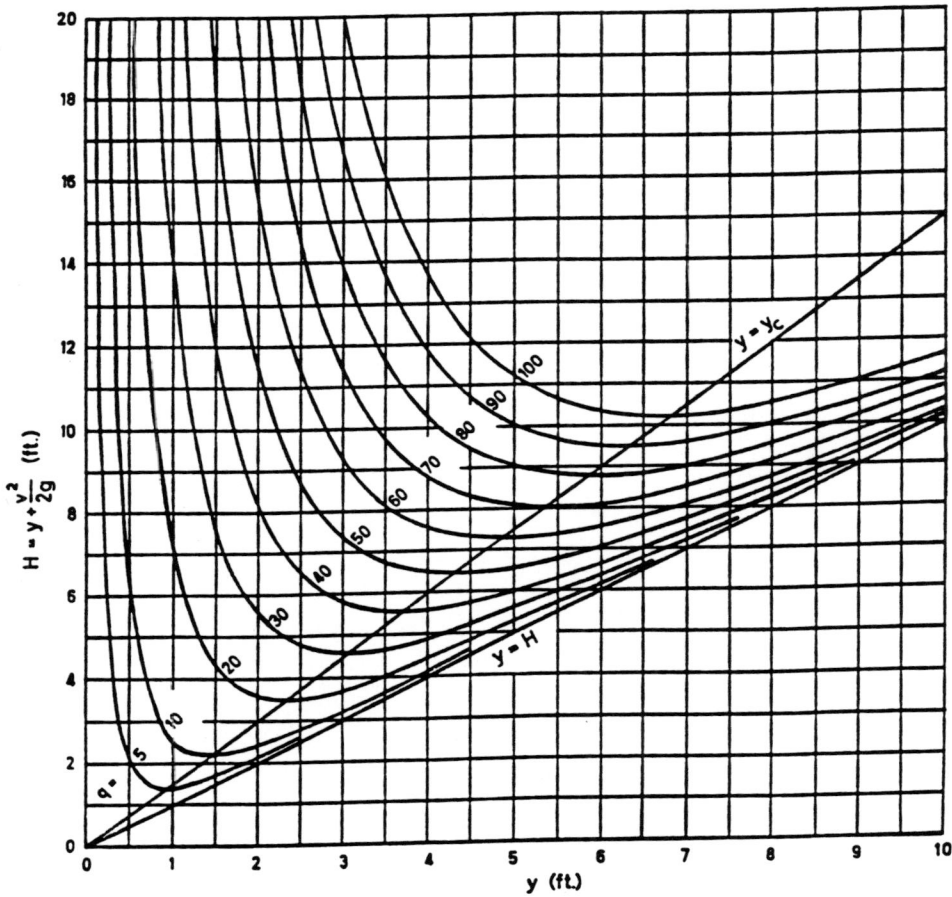

Fig. 3—Energy of flow diagram

117

$$\frac{y_2}{y_1} = \frac{1}{2}\left(\sqrt{1+8\,F_1^2}\ -\ 1\right) \qquad (5)$$

The head loss in the jump can be shown analytically to be

$$H_L = \frac{(y_2 - y_1)^3}{4y_2y_1} \qquad (6)$$

Four methods can be used for solving the above equation in the design of hydraulic structures. In the first method, the elevation of the stilling basin is initially assumed to be the same as that of the downstream channel bed; so that the energy of the flow at the toe of the spillway is

$$E_1 = y_1 + \frac{q^2}{2gy_1^2} = H + H_L \qquad (7)$$

For known values of H, H_L and q, equation (7) is solved by trial for y_1, the conjugate depth y_2 is obtained from equation (4), and the elevation of the stilling basin is determined by subtracting y_2 from the tailwater level. Apart from being approximate, this solution involves lengthly trial and error procedure which makes the method inconvenient.

One way of avoiding the tedious calculation is to use the hydraulic charts developed by Montagu (1934) and shown in Figs. 2 and 3. Given H_L and q, the energy of the flow E_2 can be directly read off the appropriate curve in Fig.2. The apron level is then obtained by subtracting E_2 from the downstream total energy line. The conjugate depths y_1 and y_2 are read off the energy curves in Fig.3. This method suffers from the disadvantage that for discharges intermediate between those shown on the curves the results must be interpolated. As this is a matter of some skill and can not be done with great measure of certainty, an accuracy of more than $\pm$ 10 per cent is not expected for the numerical results of this method.

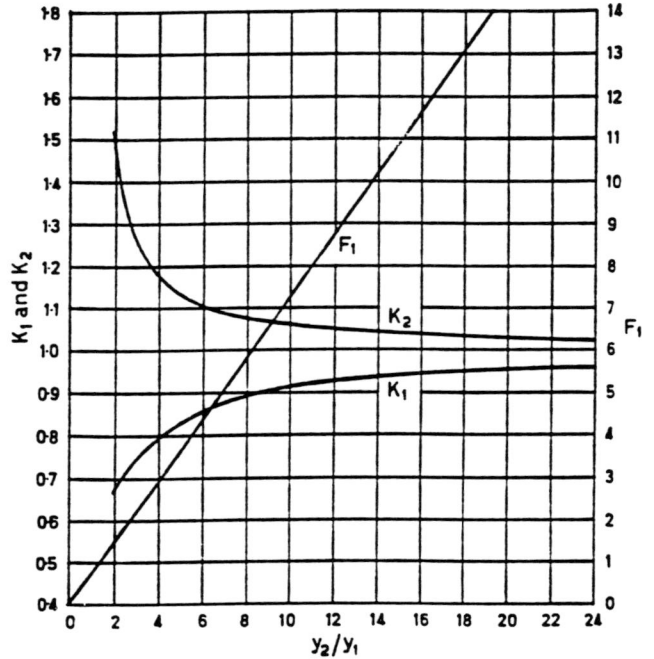

Fig. 5—Variations of K_1, K_2 and F_1

The next method is based on the rearrangement of equations (2) and (6) in terms of the critical depth $y_c = q^2/g$, thus

$$\frac{y_1}{y_c} = \left[\frac{2}{r(r+1)}\right]^{1/3} \qquad (8)$$

and

$$\frac{H_L}{y_c} = \frac{(r-1)^3}{4r}\left[\frac{2}{r(r+1)}\right]^{1/3} \qquad (9)$$

where $r = y_2/y_1$. The graphical solutions of equations (8) and (9) together with that of equation (5) are shown in Fig.4. For the known q, y_c is first calculated and then the values of y_1, y_2 and F_1 are found using this chart. The elevation of the stilling basin is determined as in the first method.

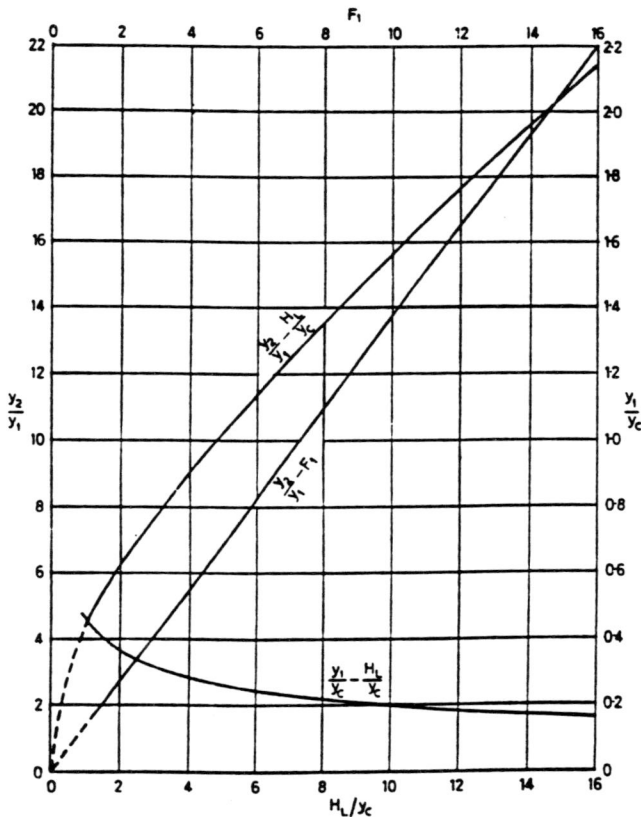

Fig. 4· Curves to determine the jump characteristics

The fourth method, which has been developed by the author, is a controlled trial and error method. Multiplying equations (2) and (6) and solving for y_2, we get

$$y_2 = \sqrt[4]{\frac{8}{g}} \left[\frac{r^4}{(1-r)^3 (1+r)} \right]^{1/4} q^{1/2} H_L^{1/4}$$

or

$$y_2 = 0.705 K_2 q^{1/2} H_L^{1/4} \qquad (10)$$

where g is taken as 32.2 ft/sec^2 and K_2 is the value of the term in the bracket. A similar expression for y_1 is obtained from equation (2).

$$y_1 = \frac{2}{g} \left[\frac{r}{(1+r)} \right] q^2 y_2^{-2}$$

or

$$y_1 = 0.062 K_1 q^2 y_2^2$$

Substituting for y_2 from equation (10).

$$y_1 = 0.125 \frac{K_1}{K_2^2} q H_L^{-\frac{1}{2}} \qquad (11)$$

The coefficients K_1 and K_2 are plotted against y_2/y_1 in Fig.5. Here, it can be seen that over the practical range $4 < y_2/y_1 < 20$ the variations in values of K_1 and K_2 are small. For design purposes, a mean value for K_1 of 0.87 and K_2 of 1.08 may be initially assumed. Having found y_1 and y_2 more accurate values of K_1 and K_2 are obtained from Fig.5 and the values of y_1 and y_2 are accordingly adjusted. Usually one such trial is sufficient to produce results within $\pm$ 4 percent.

Whatever method of design is used, it is essential that the basin elevation is found for all possible discharges through the structure, and the lowest level should be used in fixing the apron elevation.

Stilling Basin Length and Appurtenances

The design of the stilling basin is completed by determining its length and the dimensions of the appurtenances, including the glacis blocks, baffle piers and the end sill. However, these can not be calculated theoretically and recourse must be made to experiments. Recently, the U.S. Bureau of Reclamation carried out extensive field and laboratory tests and has developed generalised designs in terms of y_1, y_2 and F_1 for seven types of stilling basins (Bradley and Peterka, 1957; Elevatorsky, 1959). The reader is referred to these publications for full information.

References

1. Bradley, J.N. and Peterka, A.J., 1957. "The Hydraulic Design of Stilling Basins", A.S.C.E., Hyd. Div., Vol.83, Nos.1401 to 1406.

2. Elevatorsky, E.A., 1959. "Hydraulic Energy Dissipators", McGraw-Hill Book Co., N.Y.

3. Montagu, A.M.R., 1934. "The Standing Wave or Hydraulic Jump", C.B.I. Publ. No.7, India. See Also C.B.I. Publ. No.12, 1936.

TALUS PROTECTION WORKS

Introduction

For economic reasons, the modern trend in the design of hydraulic energy dissipators for control structures is to make the length of the stilling basin much less than the length of the hydraulic jump which forms on its apron by incorporating appurtenances which help to dissipate more energy in the jump and by providing additional protection in the form of heavy blocks overlying an inverted filter downstream of the stilling basin, Fig.1. The function of this talus is to prevent erosion of the channel bed by action of the stream and undermining by seepage flow.

The graded filter permits free movement of water but prevents the movement of sand. It consists of a layer of coarse sand followed by ballast of increasing grades and finally covered by heavy concrete blocks with open joints.

In practice, the problem is to determine the size of the block which cannot be moved by the water stream.

Most of the relevant publications on the subject deal with the stability of particles and prisms placed on bed of a uniform stream. Groat (vide Leliavsky[1]) analysed the motion of stone particles by sliding and derived an expression for the size of the block in terms of the local velocity head of the flow. Butcher and Atkinson[2] carried out model experiments to find the size of cubical blocks for preventing scour downstream of the Sennar barrage on the Nile, and correlated their results with the distance from the barrage. Allen[3] investigated over a wide range of conditions the stability of a single cube and a group of cubes forming a moud in an open channel. In the analysis he took into account the depth of flow and obtained

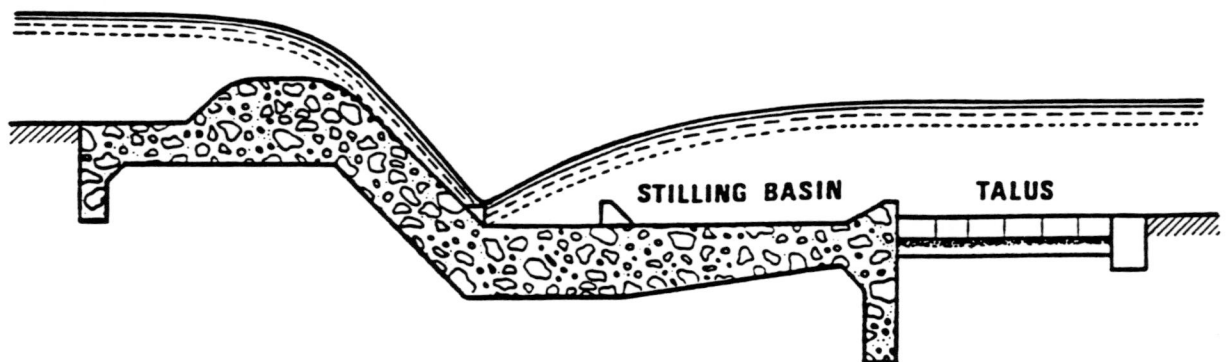

Fig. 1. Flow through a hydraulic drop structure

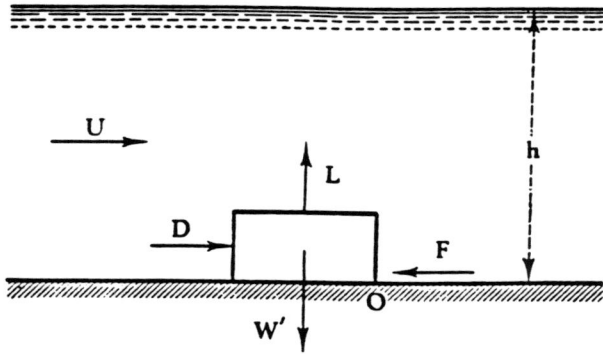

Fig. 2. Forces on a prism resting on bed of a stream

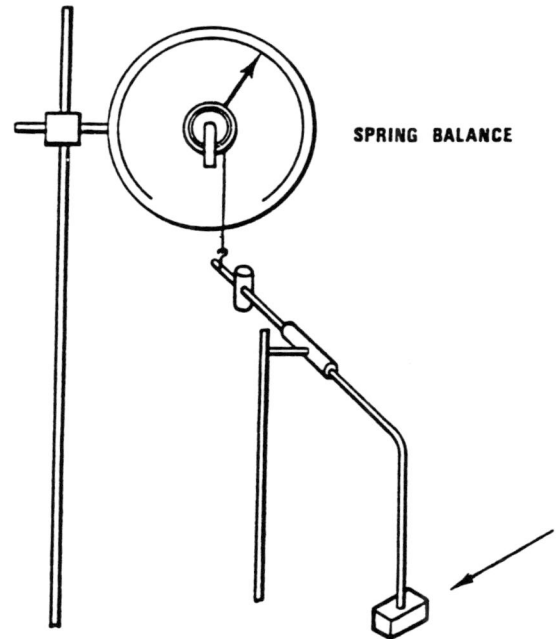

Fig. 4. Arrangement of spring balance for measuring lift force

expressions for the critical mean velocity in the section containing the prism. Pavel[4] carried out a similar investigation for stability of a prism against overturning and derived formulae for the critical mean velocity of the flow. Also he found that the stability of a row of cubes placed perpendicular to the stream is smaller than the stability of a single cube. Goncharov (vide Pavel[4]) investigated the pressures acting on cubes in a stream of water and found that the pressure acting on the front face was always positive, the pressure acting on the rear face negative. Further, if there is a row of cubes in the direction of flow, the influence of the front cube on the pressures acting on the rear cube ceases where the gap is bigger than twelve times the length of the cube.

In the present article the equilibrium of a prism on the bed of a stream is discussed for sliding and overturning in the light of lift and drag theory. Experimental results are given for the variations of lift and drag coefficients with the Froude number of the flow and for the deceleration of the hydraulic jump jet under different degrees of submergence. Using these results an equation for the size of talus blocks is derived in terms of the head loss and unit discharge.

Equilibrium of a Prism on a Stream Bed

Consider a prism of specific weight γ_1 resting on bed of a stream with mean velocity U, Fig.2. As the flow passes the block the streamlines are deflected upward to form a wake downstream. Due to the change in flow pattern, the prism is subjected to a drag force D in the direction of flow and a lift force L perpendicular to it. These hydrodynamic forces are commonly expressed by the following equations:

$$D = C_D \, \gamma \, b \, d \, \frac{U^2}{2g} \qquad (1)$$

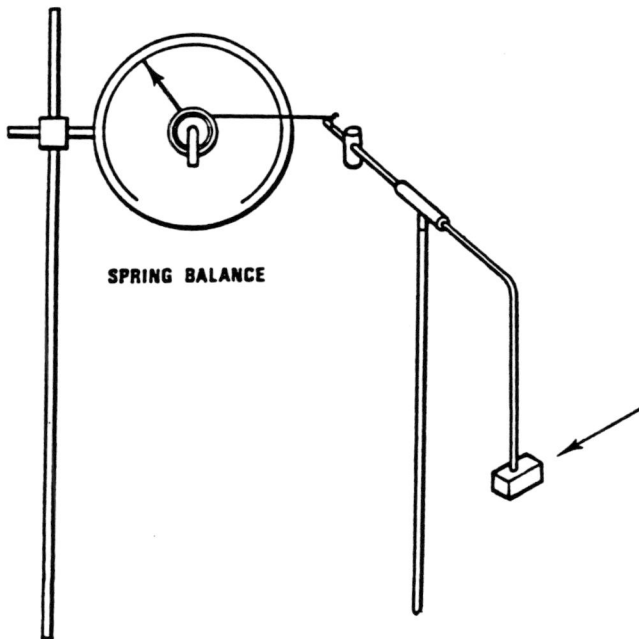

Fig. 3 Arrangement of spring balance for measuring drag force

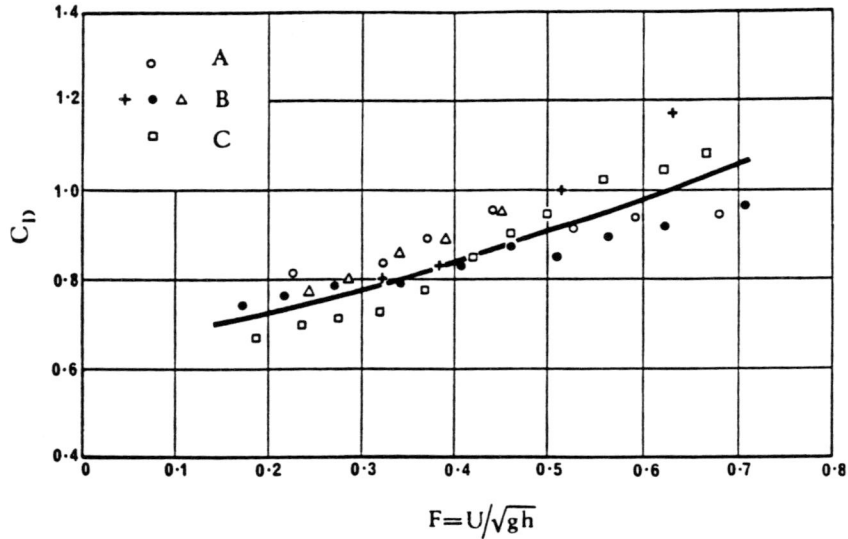

Fig. 5. Variation of drag coefficient with Froude number

$$L = C_L \, \gamma \, b \, l \, \frac{U^2}{2g} \qquad (2)$$

where b = width of prism perpendicular to flow,
d = depth of prism,
l = length of prism parallel to flow,
γ = specific weight of water,
C_D = coefficient of drag,
C_L = coefficient of lift.

The drag force is composed of skin friction and a form drag due to the pressure difference in the direction of flow. The lift force is the resultant pressure difference between the bottom and top faces of the prism. On the bottom face, since the fluid is practically stationary, the pressure is hydrostatic. On the top face the pressure is reduced below the bottom pressure partly due to the height of the prism and partly due to separation and increased velocity as a result of curvature of the streamlines.

The forces resisting the motion of the prism are the submerged weight W^1 and the sliding friction force F between the prism and the channel bed:

$$W' = b \, d \, l \, (\gamma_1 - \gamma) \qquad (3)$$

$$F = \mu \, (W' - L) = \mu \, b \, d \, l \, (\gamma_1 - \gamma)$$
$$- \mu \, C_L \, b \, l \, \frac{U^2}{2g} \qquad (4)$$

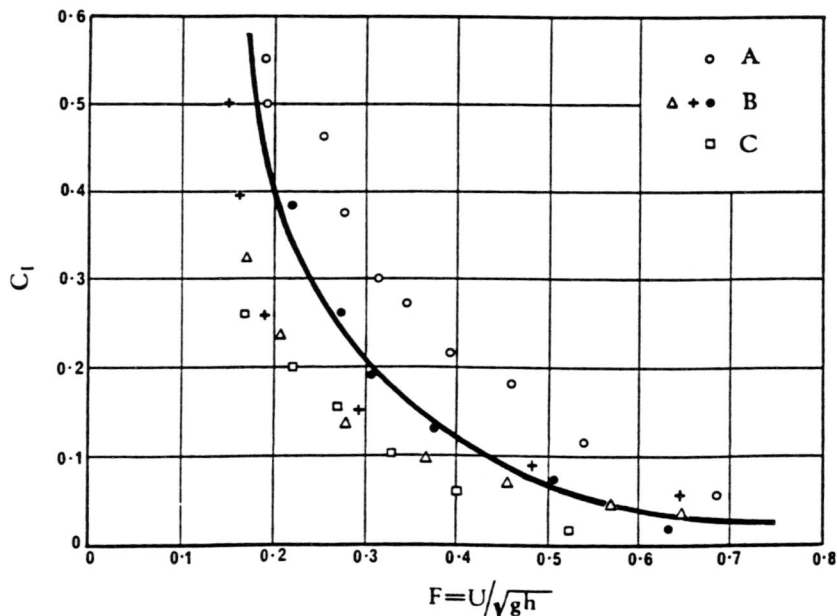

Fig. 6. Variation of lift coefficient with Froude number

122

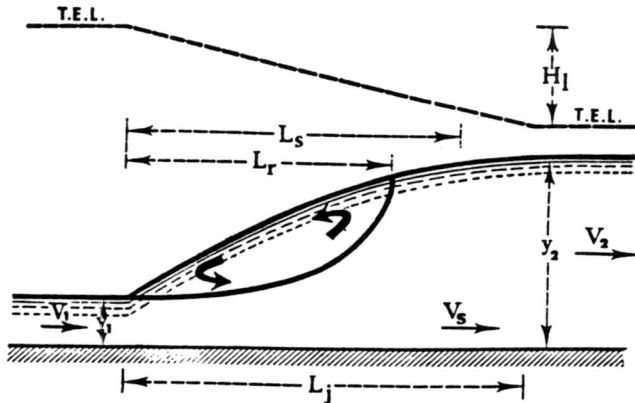

Fig. 7. Hydraulic jump on a horizontal bed

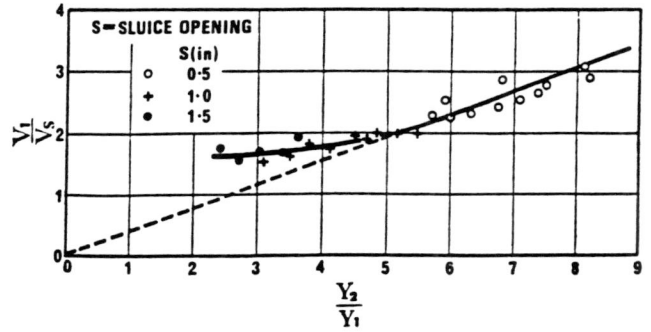

Fig. 8 Variation of velocity ratio V_1/V_s

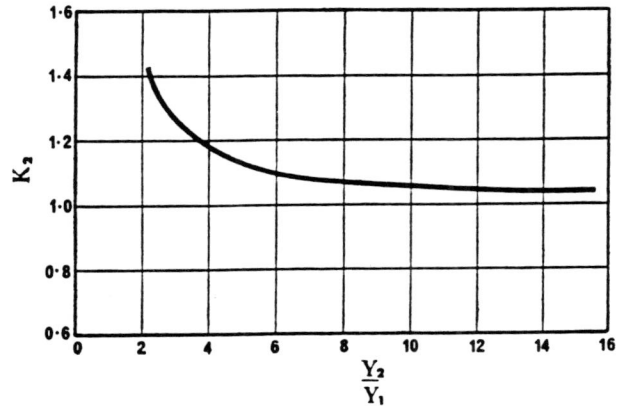

Fig. 9. Variation of coefficient K_2

μ being the coefficient of friction. The movement of the prism will depend upon the relative magnitudes and interaction of the above forces and may occur either by sliding or by overturning and rolling if initially it rests against a projection in the bed.

For equilibrium against movement by sliding, the drag force must be resisted by the sliding friction force.

$$C_D \, \gamma \, b \, d \, \frac{U^2}{2g} = \mu \, b \, d \, l \, (\gamma_1 - \gamma)$$
$$- \mu C_L \, \gamma \, b \, l \, \frac{U^2}{2g} \quad (5)$$

which reduces to

$$U = \left[\frac{\mu}{C_D + \mu \, C_L \, \frac{1}{d}} \right]^{\frac{1}{2}} \sqrt{2g \left(\frac{\gamma_1}{\gamma} - 1 \right) l} \quad (6$$

Friction tests carried out by Allen[3] by dragging cement blocks along the bed of a concrete flume with cord and spring balance indicated an average value for μ of 0.62.

For equilibrium against movement by overturning and rolling along the bed, the sum of the movements of lift and drag forces about a line through the edge 0 must equal the moment of the submerged weight about the same line. That is

$$C_D \, \gamma b d \, \frac{U^2}{2g} \, \frac{d}{2} + C_L \, \gamma b l \, \frac{U^2}{2g}$$
$$\frac{l}{2} = b d l \, (\gamma_1 - \gamma_\gamma) \, \frac{l}{2} \quad (7)$$

which gives

$$U = \left[C_D \, \frac{d}{l} + C_L \, \frac{1}{d} \right]^{-\frac{1}{2}} \sqrt{2g \left(\frac{\gamma_1}{\gamma} - 1 \right) l} \quad (8$$

Measurements of Lift and Drag Forces

To find the values of C_D and C_L, experiments were conducted in a 1ft-wide glass flume with a weighing arrangement for measuring the discharge. The drag and lift forces on plasticine prisms held about 1mm above the channel bed were measured by the torsion spring-balance arrangements shown in Figs.3 and 4. Prior to each measurement, the prism was counter-balanced by the moveable weight on the other side of the beam. The pivoted pointer attached to the spiral spring gave the force in gmf. Three sizes of prisms were used:

Mark	Dimensions
A	1.0in x 1.0in x 0.25in deep
B	1.0in x 1.0in x 0.50in deep
C	1.5in x 1.5in x 0.50in deep

Measurements were taken for various discharges and depths ranging from 4 to 16d. Values of C_D and C_L, calulated from Eqs. (1) and (2), are plotted against the

123

Froude number $F = U/(gh)^{1/2}$ in Figs.5 and 6. For depths $h < 4d$, the value of C_D was found to increase rapidly reaching 2.0 for $h = 2.3d$. Over the same range, the value of C_L was practically zero.

Deceleration of the Hydraulic Jump Jet

The purpose of a stilling basin is to dissipate the energy of the high-velocity jet leaving the weir glacis through a hydraulic jump. The physical aspects of the diffusion of this phenomenon as it occurs on a horizontal bed (Fig.7) have some similarity with the problem of the dispersion of turbulent jets and streams[5,6]. The jet expands gradually as a result of the development of a train of large-scale eddies which extend from the top of the jet to the free surface[7,8]. The mean motion of these vortices forms a surface circulation zone which is usually termed a "roller". Meanwhile, the mixing process causes the jet to expand until its upper mean flow line reaches the free surface. This point is the stagnation point which defines the downstream end of the roller. Further downstream, the depth gradually rises to the tailwater level y_2.

The length of the jump L_j is defined as the distance from the front of the jump to the point where the depth is within a few per cent of the downstream level. Bakhmeteff and Matzke[9] established experimentally that the length of the jump is six times its height,

$$L_j \simeq 6.0 \ (y_2 - y_1) \qquad (9)$$

Recent experiments by Rouse, on an air model of hydraulic jump show that the length of the roller is

$$L_r \simeq 3.5 \ (y_2 - y_1) \qquad (10)$$

For the design of dam spillways and drop structures where the jet velocity does not exceed 50ft/sec and the Froude number $F_1 = V_1/\sqrt{g \ y_1}$ is above 4.5, the length of the stilling basin is taken as[11,12]

$$L_s = 4.5 \ (y_2 - y_1) \qquad (11)$$

To calculate the size of the talus blocks, the maximum velocity V_s at distance L_s must be determined. Since this cannot be predicted theoretically, experiments were conducted in the 119ft-wide flume using a sluice gate with a rounded edge to produce a hydraulic jump. Irregularities in the supply discharge were reduced by a cylindrical overflow weir in the inlet tank. The downstream level was adjusted by an outlet weir to give different ratios of the conjugate depths y_2/y_1 ranging from 2 to 9. The velocities V_1 and V_s were measured by a Prandtl pitot-static probe. Th results are plotted non-dimensionally in Fig.8, where it is seen that for well developed jumps with $y_2/y_1 > 4$ the points lie close to a straight line passing through the origin which is represented by the following equation:

$$\frac{V_1}{V_s} = 0.385 \ \frac{y_2}{y_1} \qquad (12)$$

Formula for Size of Talus Block

Experience of the design of canal falls and regulators indicates that over a wide range of discharges and head losses the downstream Froude number of the jump $F_2 = V_2/(gy_2)^{0.5}$ lies between 0.20 and 0.30 and mostly around 0.25. Assuming this average values and square blocks with $l/d = 2.0$, from Figs.5 and 6 we obtain $C_D = 0.75$ and $C_L = 0.28$. Substituting these values in Eqs. (6) and (8), we get the mean critical velocity for sliding,

$$U = 0.75 \ \sqrt{2g \left(\frac{\gamma^1}{\gamma} - 1\right) l} \qquad (13)$$

and for overturning,

$$U = 1.035 \ \sqrt{2g \left(\frac{\gamma^1}{\gamma} - 1\right) l} \qquad (14)$$

Comparing the last two expressions it is evident that the stream velocity at which the stability of a prism is disturbed is smaller for sliding than for overturning; consequently the design of talus blocks should be based on the former condition. For concrete $\gamma^1/\gamma = 2.4$ and taking $g = 32.2$ ft/sec^2, Eq. (13) becomes

$$U = 7.14 \ \sqrt{l} \qquad (15)$$

Putting $U = V_s$ and combining Eqs. (12) and (15), we obtain

$$7.14 \ \sqrt{l} = 2.245 \ \frac{V_1 y_1}{y_2} \qquad (16)$$

Substitution of the continuity relationship:

$$q = V_1 \ y_1 \qquad (17)$$

yields the expression

$$l = 0.099 \left(\frac{q}{y_2}\right)^2 \qquad (18)$$

Using the momentum principle, the author[13] has derived the following equation for the downstream depth of the jump

$$y_2 = K_2 \left(\sqrt[4]{\frac{8}{g}}\right) q^{\frac{1}{2}} H_L^{\frac{1}{4}} \qquad (19)$$

where

$$K_2 = \left[\left(1 - \frac{y_1}{y_2}\right)^3 \left(1 + \frac{y_1}{y_2}\right)\right]^{-\frac{1}{4}} \qquad (20)$$

The coefficient K_2 is plotted against y_2/y_1 in Fig.9. Here it can be seen that over the practical range $4 < y_2/y_1 < 16$ the variation in value of K_2 is small. For design purposes a mean value for K_2 of 1.10 may be assumed and Eq. (19) then becomes

$$y_2 = 0.775 \, q^{\frac{1}{2}} \, H_L^{\frac{1}{4}} \qquad (21)$$

Substituting for y_2 in Eq. (18) we get

$$1 = 0.165 \, q \, H_L^{-\frac{1}{2}} \qquad (22)$$

For any given unit discharge q and head loss H_L, Eq. (22) determines the stable length of talus blocks downstream of the stilling basin.

References

1. Leliavsky, S. "Irrigation and Hydraulic Design", Vol.1, Chapman and Hall Ltd., London, 1955.
2. Butcher, A. and Atkinson, J. "The Causes and Prevention of Bed Erosion with Special Reference to the Protection of Structures Controlling Rivers and Canals", Min. Proc. Inst. C.E., Vol.235, Session 1932-33, Pt.I.
3. Allen, J. "An Investigation of the Stability of Bed Materials in a Stream of Water", J.Inst. C.E., No.5, p.1, March 1942.
4. Pavel, N. "Experimental and Theoretical Investigation of the Stability of Prisms on the Bottom of a Flume", Report to the 2nd Meeting International Assoc. for Hyd. Structures Research, Stockholm, Appendix 4, 1948.
5. Abramovich, G.K. "The Theory of Turbulent Jets", M.I.T. Press, Boston, Mass., U.S.A., 1963.
6. Naib, S.K.A. "Mixing of a Subcritical Stream in a Rectangular Channel Expansion", J. Inst. Wat. E., Vol.20, No.3, p.199, May 1966.
7. Naib, S.K.A. "Flow Patterns in a Submerged Liquid Jet Diffusing Under Gravity", Nature, p.694, 14 May 1966.
8. Naib, S.K.A. "Photographic Method for Measuring Velocity Profiles in a Liquid Jet", The Engineer, London, Vol.221, p.961, 24 June 1966.
9. Bakhmeteff, B.A. and Matzke, A.E. "The Hydraulic Jump in Terms of Dynamic Similarity", Trans. ASCE, Vol.101, p.630, 1936.
10. Rouse, H. Siao, T.T. and Nagaratnam, S. "Turbulence Characteristics of the Hydraulic Jump", Journ. Hyd. Div., Proc. ASCE, Vol.84, Part I, p.1528, 1958.
11. Bradley, J.N. and Peterka, A.J. "The Hydraulic Design of Stilling Basins", Journ. Hydr. Div., Proc. ASCE, Vol.83, Nos. 1401 to 1406, 1957.
12. Elevatorsky, E.A. "Hydraulic Energy Dissipators", McGray-Hill Book Co., New York, 1959.
13. Naib, S.K.A. "Hydraulic Design of Energy Dissipators", Water and Water Engineering, London, Vol.70, p.191, May 1966.

PART IX

HYDRAULIC CONTROL STRUCTURES

Hydraulic Drop Structures
Uplift Under Hydraulic Structures

HYDRAULIC DROP STRUCTURES

Introduction

Canal falls are provided in irrigation schemes to maintain a certain required drop in canal level through virtually the whole range of operating discharges. In the case of a regulated fall, it also serves to obtain command at specific points along a canal, where the water surface has to be raised to enable an offtaking main or branch canal to be supplied. In both cases, a drop in water level is artificially created which results in the conversion of potential head to kinetic head. This head must be dissipated within the boundaries of the structure or it will destroy the downstream channel by erosion, thus endangering the safety of the structure.

During the past fifty years, various research stations have carried out design studies on canal falls. The Indian Waterways Experimental Station developed in 1944 an effective design for a flume meter "Baffle Fall". However, since this type of fall is a metering weir, it requires a very long flat glacis and long apron which make the design costly. Subsequently, the Central Water and Power Research Station at Poona carried out a programme of ad hoc research and model experiments for different designs of falls under various conditions of discharge and drop. A survey of both these developments was reported by Talwani et al [1].

In 1955, the U.S. Bureau of Reclamation carried out extensive field and laboratory tests and developed generalised designs for short stilling basins of dams and drop structures [2]. These designs are widely used but appropriate protection works must be provided downstream to suit local conditions. In a model study of one type of design, appreciable erosion was observed on the sides and bed of the downstream channel and it was necessary to provide a suitable length of protection downstream of the wingwalls.

As a further contribution, the author has carried out experimental and theoretical investigations in order to analyze the flow through a control drop structure and to evolve a generalised procedure of design for a talus type canal fall for irrigation works. To this end, a study of the flow over the weir of the drop was undertaken with model tests to develop a simplified weir shape with known discharge characteristics. Methods of calculating the elevation of the stilling basin were then presented [3]. This was followed by investigations of the expansions of flows in rectangular and trapezoidal channel expansions and their application to the design of protection downstream of the stilling basin [4] [5]. This work was extended by studying the transport process of a prism on the bed of a stream and methods of analyzing its equilibrium are discussed in terms of lift and drag concepts. Based on this theory and experimental measurements of the deceleration of the hydraulic jump jet, an engineering design equation was derived for the size of talus blocks downstream of the stilling basin [6].

This paper presents the research results in a form readily usable by engineers for the design of hydraulic drop structures and irrigation canal falls. The work forms part of the current research on the diffusion of confined turbulent jets and stream and their application to the design of hydraulic structures, selected recent references are given [7] [8].

Talus Type Canal Falls

The canal fall developed in this paper is a talus type structure, Figure 1. The basic shape has been evolved to standardize the procedure of calculations and to reduce the necessary hydraulic research to aspects of one model. The fall consists of an upstream protection in the form of stone pitching, an essentially curved weir with a short flat crest providing seating for gates in regulated falls, joined to a 45° inclined glacis, ending in a horizontal stilling basin

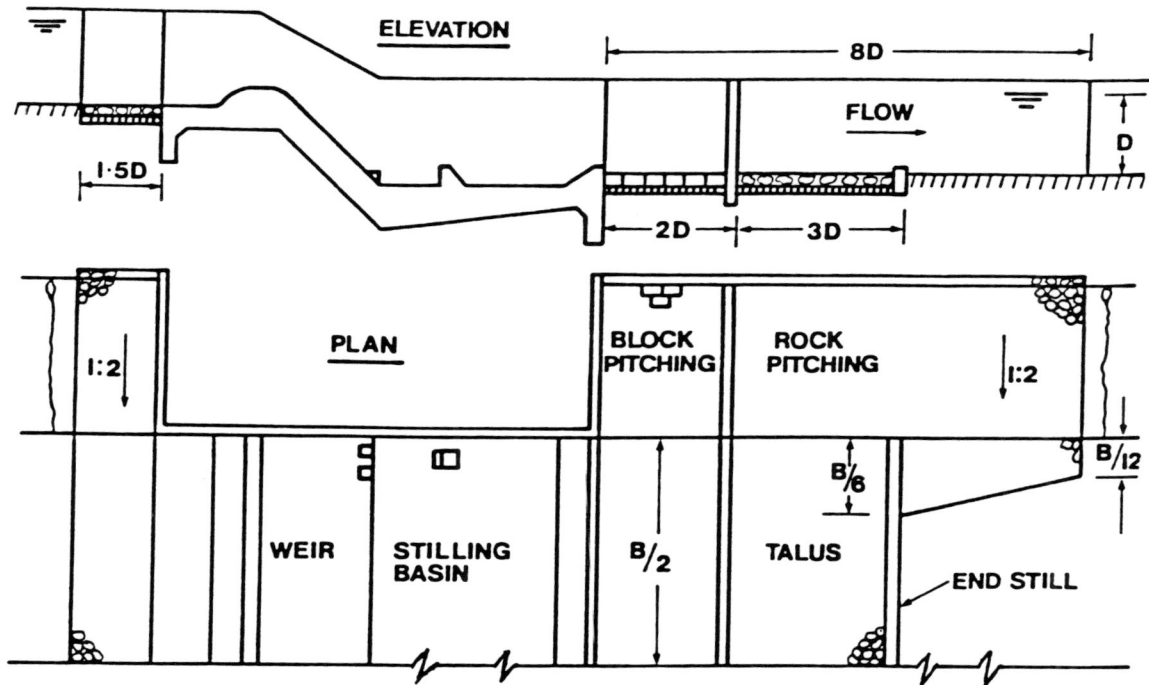

Figure 1 – Talus Type Canal Fall

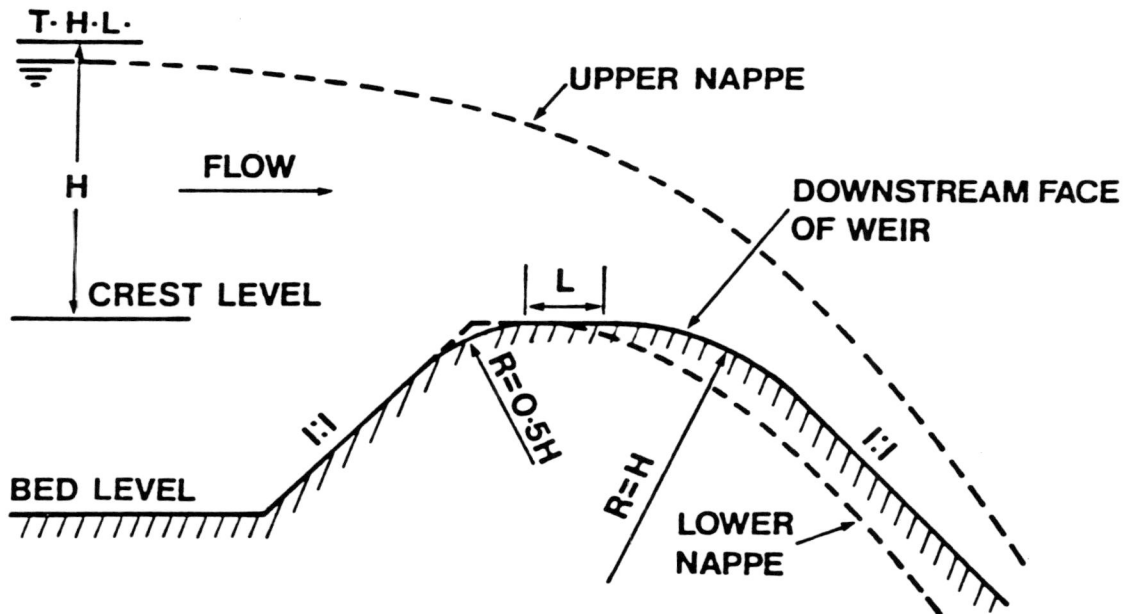

Figure 2 – Standard Weir Shape

with appurtenances to intensify energy dissipation in the hydraulic jump occurring under designed conditions. This is followed by a talus protection consisting of concrete blocks overlying an inverted filter and stone pitching ending with a lip wall on the bed further downstream. The reverse filter acts as a safeguard against excessive uplift pressure downstream of the stilling basin.

Weir Shape and Model Studies

The development of a basic weir shape is required to simplify the design and construction of a large number of falls in a single irrigation scheme and also to economise on any hydraulic laboratory testing. The co-ordinates of the weir were initially calculated by multiplying by the specific head H the values for the lower and upper nappes of a standard dam spillway with the face inclined on a 1:1 slope [9]. The lower nappe was then substituted as shown in Figure 2, by two arcs with radii 0.5H and H and an intermediate straight length L to be sufficient for the seating of gates in the case of regulated falls. The curves join the upstream and downstream faces of the weir as tangents to avoid separation of the jet and negative pressures on the weir face.

Based on experience of using 4m span radial gates on falls with capacities up to 100^3/s and bed width up to 30m, the values of the straight lengths taken for various unit discharges q are given in Table 1.

q (m³/s/m)	1.5-2.0	2.0-2.5	2.5-3.0	3.0-3.5
L (m)	0.4	0.5	0.6	0.7

Table 1 - Variation of L with unit discharge q

The discharge formula relating to weirs is:

$$Q = CB(H)^{\frac{3}{2}}$$

Where Q is the total discharge, B is the length of the weir, H is the head above the crest and C is the coefficient of discharge.

Tests were carried out in a glass flume of a 1/20 model of a 3m fall and a 1/15 model of a 0.9m fall, both constructed in timber with a varnished finish of all surfaces. The water supply to each model was provided from a constant head tank and measured by a weighing machine. Upstream and downstream water levels were recorded by a point gauge. An adjustable weir at the end of the flume controlled the downstream water levels. The calibration for each fall under free discharge conditions was undertaken by relating the upstream head to the

corresponding discharge through the structure. For each discharge the downstream water level was adjusted to the downstream level from the rating curve of the particular fall.

The variation of C for the two falls are plotted against the ratio of discharge to the design discharge Q/Q_D in Figure 3, where it is seen that the points lie close to a mean curve. This curve is believed to be applicable to any fall with similar crest design.

Theory and Design of Stilling Basins

Consider the flow in a channel drop structure shown in Figure 4. In practice the upstream and downstream bed levels, the depth of flow D, the head loss H_L and the unit discharge q are known. It is required to calculate y_1, y_2 and the elevation of the stilling basin. Applying the continuity, momentum and specific head equations the following expressions may be derived:

$$y_1 y_2 (y_1 + y_2) = \frac{2}{g} q^2 \qquad (1)$$

$$y_1 = \frac{-y_2}{2} + \left(\frac{2q^2}{gy_2} + \frac{y_2^2}{4} \right)^{\frac{1}{2}} \qquad (2)$$

$$y_2 = \frac{-y_1}{2} + \left(\frac{2q^2}{gy_1} + \frac{y_1^2}{4} \right)^{\frac{1}{2}} \qquad (3)$$

$$H_L = \frac{(y_2 - y_1)^3}{4 y_2 y_1} \qquad (4)$$

$$E_1 = y_1 + \frac{q^2}{2gy_1^2} = H + H_L \qquad (5)$$

Four methods can be used for solving the above equations for design purposes. In the first method, the elevation of the stilling basin is initially assumed to be the same as that of the downstream channel bed. For known values of H, H_L and q, equation (5) is solved by trial for y_1, the conjugate depth is obtained from equation (3) and the elevation of the stilling basin is determined by subtracting y_2 from the tailwater level. This solution involves lengthy trial and error procedures which are best solved by a computer.

A second method is to use a specific head diagram for a series of unit discharges. This method suffers from the disadvantage that for the discharges intermediate between those shown on the curves the results must be interpolated. A third method is based on the rearrangement of equations

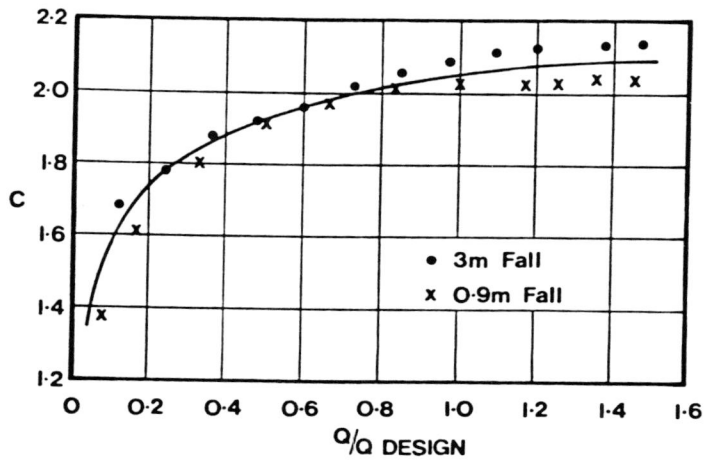

Figure 3 - Coefficient of Discharge for Weir

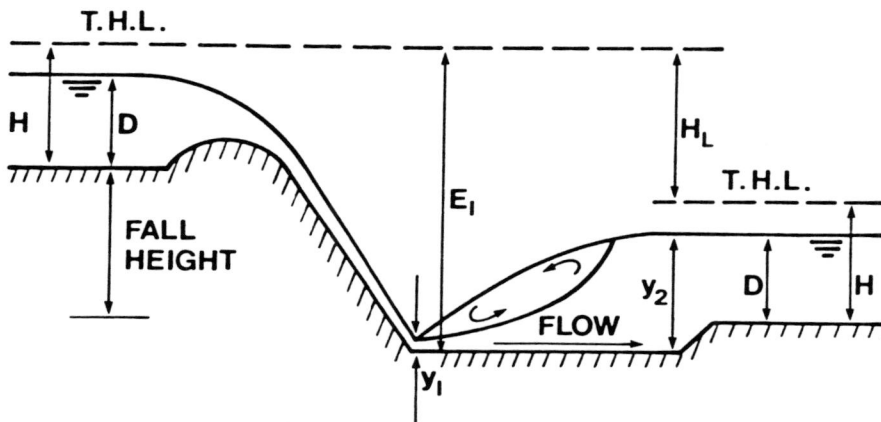

Figure 4 - Flow in a Channel Drop Structure

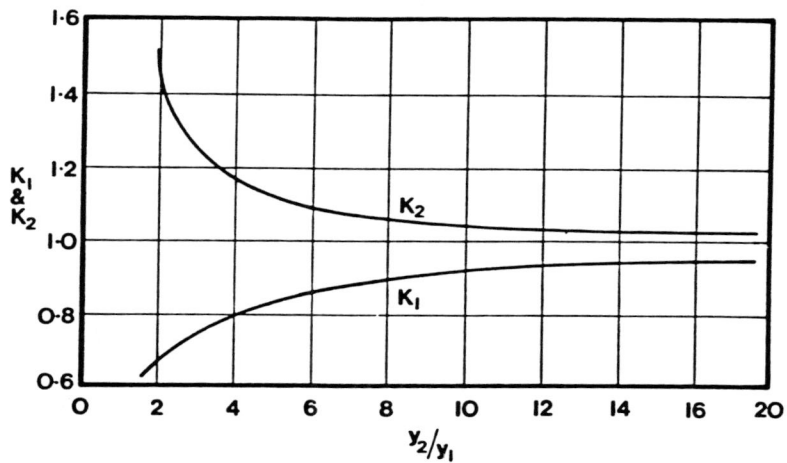

Figure 5 - Variations of K_1 and K_2

131

(1) and (4) in terms of the critical depth and the graphical solutions of the resulting equations [10].

A fourth method, developed by the author, [3] is a controlled trial and error method. Multiplying equations (1) and (4) and solving for y_1 and y_2 we get

$$y_1 = 0.23 \frac{K_1}{K_2} q H_L^{-\frac{1}{2}} \qquad (6)$$

$$y_2 = 0.95 K_2 q^{\frac{1}{2}} H_L^{\frac{3}{4}} \qquad (7)$$

$$where \ r = y_2/y_1 \ ; \ K_1 \ \frac{r}{(1+r)}$$

$$K_2 = \left(\frac{r^4}{(1-r)^3 (1+r)} \right)^{\frac{1}{4}}$$

The coefficients K_1 and K_2 are plotted against y_2/y_1 in Figure 5. Here it can be seen that over the practical range $4 < y_2/y_1 < 20$ the variations in values of K_1 and K_2 are small. For design purposes, a mean values of K_1 of 0.87 and K_2 of 1.08 may be initially assumed. Having found y_1 and y_2 more accurate values of K_1 and K_2 are obtained from Figure 5 and the values of y_1 and y_2 are accordingly adjusted. Usually one such trial is sufficient to produce results within ± 3 per cent.

Whatever method of design is used, it is essential that the basin elevation is found for all possible discharges through the structure, and the lowest level should be used in fixing the apron elevation.

The design of the stilling basin is completed by determining its length and the dimensions of the appurtenances, including the chute blocks, baffle piers and the end sill. However, these can not be calculated theoretically and recourse must be made to experiments. The U.S. Bureau of Reclamation carried out extensive field and laboratory tests and developed generalised designs in terms of y_1, y_2 and F_1 for seven types of stilling basins [2]. For the range of Froude numbers for the incoming flow between 4.5 and 9, the basin shown in Figure 6 is adopted. The leading dimensions have been checked as a result of tests carried out on a 1/20 model of a 3.0m fall.

Flow in the Trapezoidal Channel Expansion

When the stream leaves the stilling basin it is discharged abruptly into the downstream channel with a velocity normally higher than the maximum permissible value for the soil formation and if the channel is unprotected scour will occur. Furthermore, the interface between the stream and the circulation fluid consists of a series of large scale eddies which, because of their strong vortex motion, erode the channel banks and destroy their stability. To safeguard against these two actions, a protection must be provided in the form of talus blocks and stone pitching as shown in Figure 1.

To design the length of the protection, experiments were carried out on a rectangular stream expanding abruptly into a trapezoidal channel expansion [5]. At any depth in the trapezoidal expansion, the pattern of flow is shown in Figure 7. Between the uniform stream with a velocity U_1 and the adjacent slow moving fluid a turbulent mixing layer is formed having an inner boundary 01 and an outer boundary 02. The mixing process between the stream and the surrounding fluid causes a circulation zone to be established in which the reverse velocity U_2 increases with distance from JJ', until it reaches it maximum value at section MM^1 and then decreases to zero at the stagnation point. This flow resembles that for a rectangular channel expansion for which a theoretical solution was obtained [4] based on Abramovich's analysis of the flow behind a two-dimensional bluff body [11].

The experimental and theoretical results for the boundaries of the stream, measured 0.2 of the depth below the water surface are shown in Figure 8. The inner boundary is seen to expand at an angle of 5 degrees while the outer layer follows a curve, the slope of which indicates a much greater rate of spreading than that predicated for 02. At section $x/B = 3.75$, the outer boundary joins to the wall layer on the side of the channel, thus extending the turbulent mixing layer across the full width of the circulation zone.

Figure 9 shows the variation in velocity of the backflowing stream, together with the theoretical and experimental results for the rectangular expansion. The maximum reverse velocity agrees well with the theoretical value of $0.4U_1$, but it occurs at $x/B = 3.75$, the section where the outer boundary reaches the side JN. This section also defines the downstream end MM^1 of the first region of the circulation zone.

The total length of the circulation zone is 5.6B compared with the theoretical value of 6.0B and that for rectangular expansion of 6.5B. It should be noted that due to the side slope in the trapezoidal expansion the length of the circulation varies with depth, ranging from zero at the bed to a maximum at the water surface. The length of the transition protection downstream of a channel expansion should generally be fixed by the length of the circulation zone, which for a side slope of 1:2, is about eleven times the depth of flow. Due to the greater decay of the maximum velocity downstream of the stilling basin, model investigations show that the length of the talus protection can be reduced to eight times the depth.

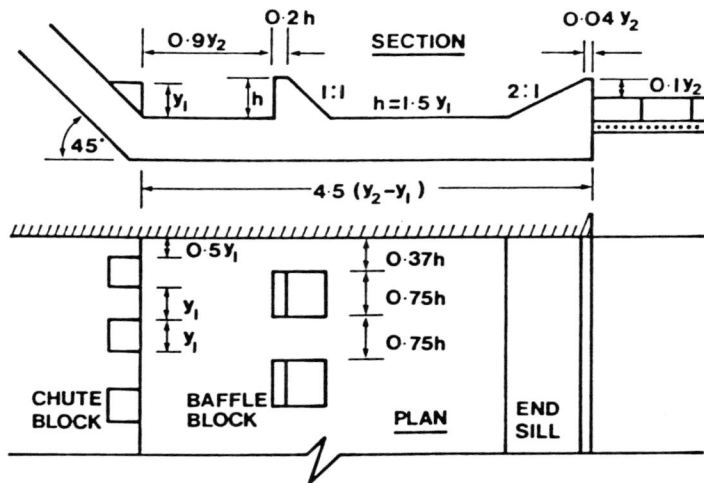

Figure 6 - Stilling Basin Appurtenances

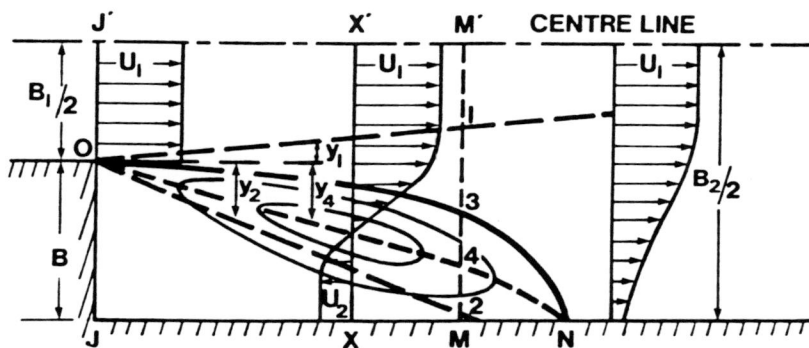

Figure 7 - Flow in a Trapezoidal Channel Expansion

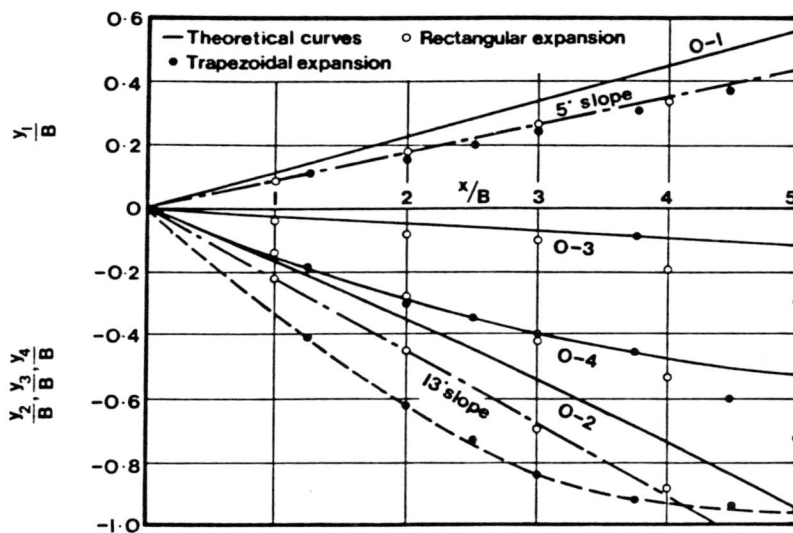

Figure 8 - Spreading of Stream Boundaries

Equilibrium of a Prism on a Stream Bed

To calculate the size of the talus blocks, consider a prism of density ρ_1, length l, depth d and width b, resting on a bed of a stream with a mean velocity U. As the flow passes the block the streamlines are deflected upward to form a wake downstream. Due to the change in flow pattern, the prism is subjected to a drag force D in the direction of flow and a lift force L perpendicular to it. These forces are commonly expressed as follows:

$$D = C_D \ \rho g b d \left(\frac{U^2}{2g} \right) \qquad (8)$$

$$L = C_L \ \rho g b l \left(\frac{U^2}{2g} \right) \qquad (9)$$

here ρ is the density of water, C_D is the drag coefficient and C_L is the lift coefficient.

The forces resisting the motion of the prism are the submerged weight W^1 and the sliding friction force F between the prism and the channel bed:

$$W^1 = bdlg \ (\rho_1 - \rho) \qquad (10)$$

$$\begin{aligned} F = \mu (W^1 - L) &= \mu \ bdlg \ (\rho_1 - \rho) \\ &- \mu \ C_L^{\frac{1}{2}} \ \rho b l \ U^2 \end{aligned} \qquad (11)$$

μ being the coefficient of friction. The movement of the prism will depend upon the relative magnitudes and interaction of the above forces and may occur either by sliding or by overturning and rolling if initially it rests against a projection in the bed.

For equilibrium against movement by sliding it can be shown:

$$U = \left(C_D + C_L \ \frac{1}{d} \right)^{-\frac{1}{2}} \left(2g \left(\frac{\rho_1}{\rho} - 1 \right) 1 \right)^{\frac{1}{2}} \mu$$

$$(12)$$

For equilibrium against movement by overturning and rolling along the bed, a similar equation may be derived:

$$U = \left(C_D \ \frac{d}{1} + C_L \ \frac{1}{d} \right)^{-\frac{1}{2}} \left(2g \left(\frac{\rho_1}{\rho} - 1 \right) 1 \right)^{\frac{1}{2}}$$

$$(13)$$

Friction tests carried out by Allen [12] by dragging cement blocks along the bed of a concrete flume indicated an average values for μ of 0.62.

To find the values of C_D and C_L experiments were conducted in a 300mm glass flume with a weighing arrangement for measuring the discharge. The drag and lift forces on plasticine prisms held about 1mm above the channel bed were measured by torsion-balance arrangements [6]. Measurements were taken for various discharges and depths h ranging from 4 to 16d. Mean curves for values of C_D and C_L calculated from Equations (8) and (9) are plotted against the channel flow Froude number $F = U/(gh)^{0.5}$ in Figure 10.

Deceleration of the Hydraulic Jump Jet

The purpose of a stilling basin is to dissipate the energy of the high velocity jet leaving the weir glacis through a hydraulic jump. The physical aspects of the diffusion of this phenomenon as it occurs on a horizontal bed (Figure 11) have similarity with the problem of the dispersion of confined jets and streams. The jet expands gradually as a result of the development of a train of large scale vortices which extend from the top of the jet to the free surface. The mean motion of the vortices forms a surface circulation zone which is usually termed as a "roller". Meanwhile, the mixing process causes the jet to expand until its upper mean flow line reaches the free surface. This point is the stagnation point which defines the downstream end of the roller. Further downstream, the depth gradually rises to the tailwater level y_2.

For the design of hydraulic drop structures where the Froude number $F_1 = V_1/(gy_1)^{1/2}$ is above 4.5, the length of the stilling basin is taken as:

$$L_s = 4.5 \ (y_2 - y_1) \qquad (14)$$

To calculate the size of the talus blocks, the maximum profile velocity V_s at distance L_s must be determined. Since this cannot be predicted theoretically, experiments were initially conducted in a flume using a sluice gate. The downstream water level was adjusted to give different ratios of the conjugate depths y_2/y_1 ranging from 2 to 9. The velocities V_1 and V_s were measured by a Prandtl pitot-static probe. The results are plotted non-dimensionally in Figure 12 where it is seen that for well developed jumps with $y_2/y_1 > 4$ the points lie close to a straight line passing through the origin which is represented by the following equation:

$$\frac{V_1}{V_s} = 0.39 \left(\frac{y_2}{y_1} \right) \qquad (15)$$

134

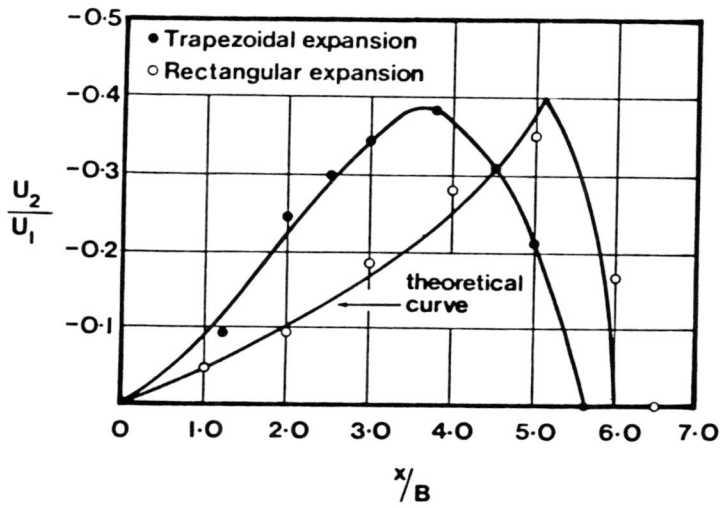

Figure 9 - Velocity of Reverse Stream

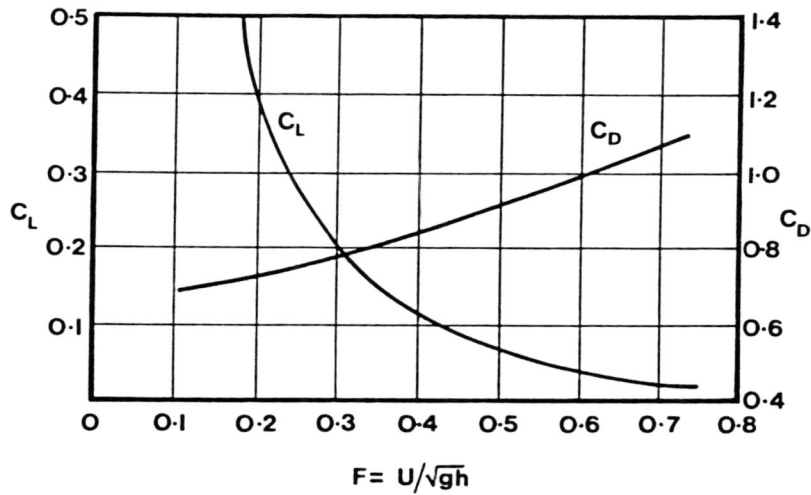

Figure 10 - Lift and Drag Coefficients C_D and C_L

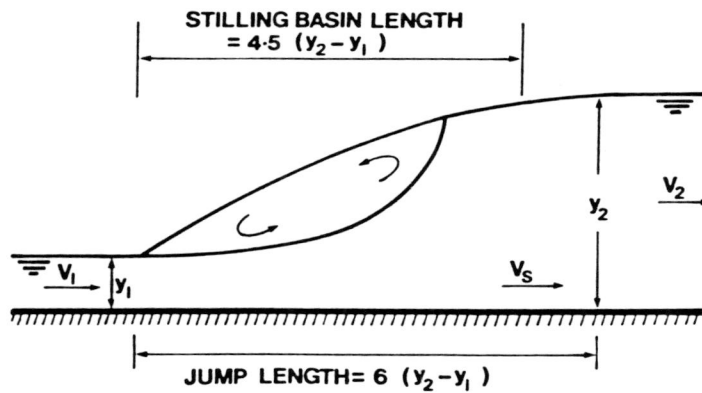

Figure 11 - Hydraulic Jump on a Horizontal Bed

The U.S.B.R. standard stilling basin relies on dissipation of energy by the chute blocks, baffle blocks and the end sill as well as on turbulence of the jump phenomena. Because of the greater energy dissipation the maximum velocity V_s downstream of the stilling basin is smaller than that indicated by equation (15) for the undisturbed jump on a horizontal floor. Measurements of maximum velocities near the bed in a 1/20 model of a 3m fall were made using a miniature current meter. The measured values are plotted in Figure 12 and the mean line drawn through them is represented by the equation:

$$\frac{V_1}{V_s} = 0.63 \left(\frac{y_2}{y_1} \right) \qquad (16)$$

Formula for size of Talus Blocks

Experience of the design of canal falls and regulators indicates that over a wide range of discharges and head losses the downstream Froude number of the jump $F_2 = V_2/(gy_2)^{0.5}$ lies between 0.20 and 0.30 and mostly around 0.25. Assuming this average value and square blocks with $1/d = 2.0$, from Figure 10, we obtain $C_D = 0.75$ and $C_L = 0.28$. Substituting these values in Equations (11) and (12) we get the mean critical velocity for sliding,

$$U = 0.75 \left(2g \left(\frac{\rho_1}{\rho} - 1 \right) 1 \right)^{\frac{1}{2}} \qquad (17)$$

and for overturning,

$$U = 1.035 \left(2g \left(\frac{\rho_1}{\rho} - 1 \right) 1 \right)^{\frac{1}{2}} \qquad (18)$$

Comparing the last two expressions it is evident that the stream velocity at which the stability of a prism is disturbed is smaller for sliding than for overturning; consequently the design of talus blocks should be based on the former condition. For concrete $\rho_1/\rho = 2.4$ taking $g = 9.81$ m/sec^2, Equation (17) becomes

$$U = 2.99 \ (1)^{\frac{1}{2}} \qquad (19)$$

Putting $U = V_s$ and combining Equations (17) and (19) we obtain

$$2.99 \ (1)^{\frac{1}{2}} = 1.59 \ \frac{V_1 y_1}{y_2} \qquad (20)$$

Substitution of the continuity relationship:

$$q = V_1 y_1 \qquad (21)$$

yields the expression

$$1 = 0.28 \left(\frac{q}{y_2} \right)^2 \qquad (22)$$

For design purposes a mean value for K_2 of 1.08 may be assumed and Equation (7) then becomes

$$y_2 = 1.02 \ q^{\frac{1}{2}} \ H_L^{\frac{1}{4}} \qquad (23)$$

Substituting for y_2 in Equation (22) we get

$$1 = 0.27 \ q \ H_L^{-\frac{1}{2}} \qquad (24)$$

For any given unit discharge q and head loss H_L, Equation determines the stable length of talus blocks downstream of the stilling basin.

References

1. Talwani, B.S. et al (1952). Design of Canal Falls. Journal of Irrigation & Power, April, 269-288.
2. Bradley, J.N. et al (1957). The Hydraulic Design of Stilling Basins. Proceedings of the American Society of Civil Engineers, Paper 1403, 1-22.
3. Naib, S.K.A. (1966). Hydraulic Design of Energy Dissipators. Water & Water Engineering, May, Vol 70, 191-193 and (1967), Vol. 71, 336.
4. Naib, S.K.A. (1966). Mixing of a Subcritical Stream in a Rectangular Channel Expansion. Journal of the Institution of Water Engineers, Vol.20, 199-206.
5. Naib, S.K.A. (1969). Stream Boundaries of Subcritical Flow in a Trapezoidal Channel Expansion. Water & Water Engineering, April, 155-158.
6. Naib, S.K.A. (1967), Equilibrium of Talus Blocks Downstream of Stilling Basins. Water Power, October, 407-410.
7. Naib, S.K.A. (1974). Deflexion of a Submerged Round Jet to Increases Lateral Spreading. La Houille Blanche, No. 6, 455-461.
8. Naib, S.K.A. (1980). Surface Motion of a Plane Liquid Jet. La Houille Blanche, No.6, 377-383.
9. Hinds, J. et al (1945). Engineering for Dams. Vol. II, 358-361.
10. Elevatorsky, E.A. (1959). Hydraulic Energy Dissipators. McGraw-Hill Book Co., 22-30.
11. Abramovich, G.K. (1963). The Theory of Turbulent Jets. M.I.T. Press, 391-443.
12. Allen, J. (1942). An Investigation of the Stability of Bed Materials in a Stream of Water. Journal of the Institution of Civil Engineers, No. 5, 1-6.

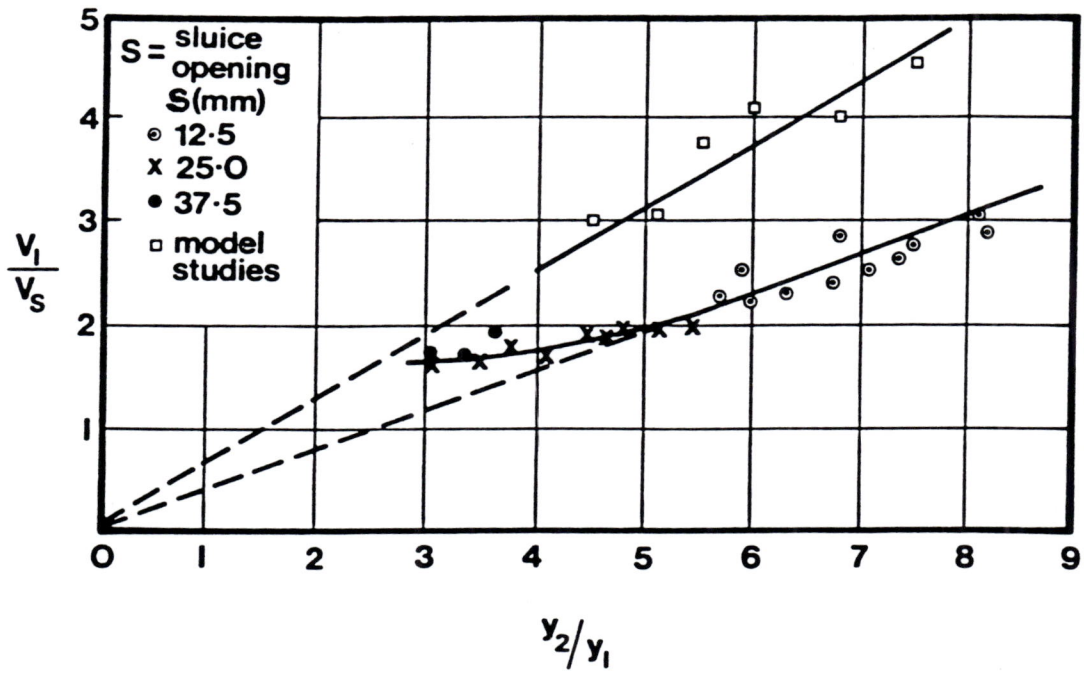

Figure 12 - Variation of Velocity Ratio V_1/V_s

UPLIFT UNDER HYDRAULIC STRUCTURES

Summary

Economical design of hydraulic control structures resting on permeable foundations requires accurate determination of the uplift pressure distribution and the exit gradient. In this article, the relevant theoretical solutions of Weaver and the method of independent variables of Khosla are briefly presented and compared with experimental methods using an electrical analogue and a seepage tank recently developed by the author.

Introduction

As water flows through the soil under a regulator or a weir, gradually it loses head due to soil resistance until at the exit the head is reduced to zero. Thus, there will be an uplift pressure acting on the underside of the structure, and a certain exit gradient at the downstream end where water emerges from the soil. If at this end the upward force of the water is greater than the submerged weight of the soil particles they will be washed away: this is a dangerous condition which eventually may lead to the failure of the structure.

It follows that there are two essential considerations in the subsoil design of a hydraulic structure. For stability of the structure against uplift, a suitable floor thickness must be provided at different sections such that the weight balances the pressure force at each section. To safeguard against undermining the exit gradient must not exceed a certain safe limit generally 0.2 to 0.15. This is achieved by providing a suitable depth of pile line or cutoff wall at the downstream end of the floor.

For simple structures, theoretical solutions can be found for the distribution of uplift pressure and the value of the exit gradient. However, for most practical designs the boundary conditions are so complex that the theoretical equations are too difficult, and often impossible to solve. To overcome this difficulty, Khosla et al[1] developed a method in which the theory for a flat floor with a pile at the downstream end is used in conjunction with empirical equations and charts. The method is known as the method of independent variables.

Besides this method, hydraulic and electrical models can be used to determine the flow lines and the distribution of uplift pressure. The electrical analogue may be conveniently used in the design office for analysing structures resting on confined permeable foundations of complex geometry. The object of this article is twofold:

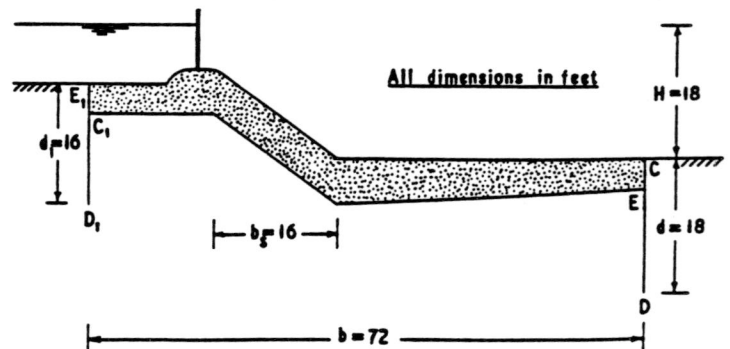

Fig. 1   Elevation of hydraulic drop structure

138

(1) to familiarise young engineers with the theory and procedure of Khosla's method, and (2) to compare the results of this method with those of the analogue and hydraulic models for a regulator and drop structure (Fig.1) in order to assess the differences involved for design purposes.

Theoretical Solutions

The fundamental equations governing two-dimensional flow of water through soils are those of Darcy and Laplace,

$$Q = kA \frac{dh}{ds}$$

$$\frac{\partial^2 \phi}{\partial x^2} + \frac{\partial^2 \phi}{\partial y^2} = 0$$

$$\frac{\partial^2 \psi}{\partial x^2} + \frac{\partial^2 \psi}{\partial y^2} = 0$$

where Q is the flow rate, ϕ the potential function and ψ the stream function.

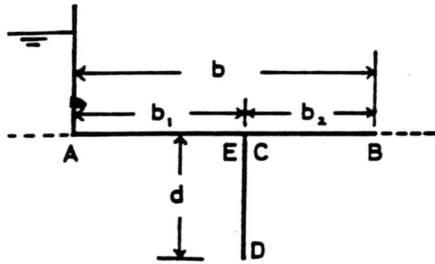

Fig. 2   Floor with an intermediate pile

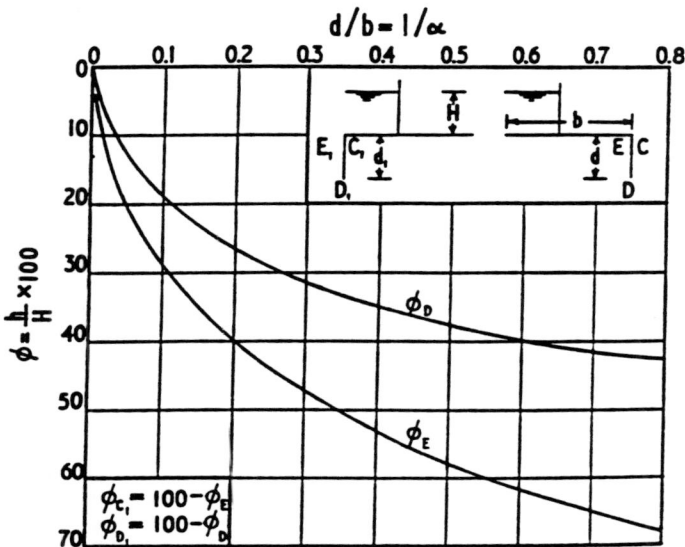

Fig. 3   Pressures for a floor with an end pile

Using conformal transformation, Weaver[2] and Khosla et al.[1] solved the above equation for the case of a horizontal floor with a vertical pile line at any point along its length resting on an infinite strata. The solutions for a floor with an intermediate pile line (Fig.2) are:

$$\phi E = \frac{1}{\pi} \cos^{-1}\left(\frac{\lambda_1 - 1}{\lambda}\right)$$

$$\phi C = \frac{1}{\pi} \cos^{-1}\left(\frac{\lambda_1 + 1}{\lambda}\right)$$

$$\phi D = \frac{1}{\pi} \cos^{-1}\left(\frac{\lambda_1}{\lambda}\right)$$

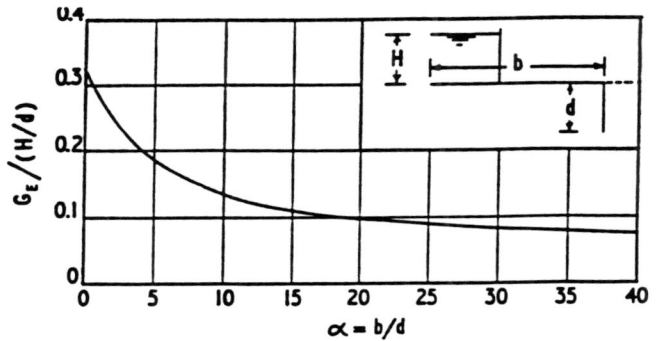

Fig. 4   Exit pressure gradient

where

$$\lambda = \frac{1}{2}\left(\sqrt{1+\alpha_1^2} + \sqrt{1+\alpha_2^2}\right)$$

$$\lambda_1 = \frac{1}{2}\left(\sqrt{1+\alpha_1^2} - \sqrt{1+\alpha_2^2}\right)$$

$$\alpha_1 = b_1/d \qquad \alpha_2 = b_2/d$$

For a floor with a pile line at end (Fig.3) the following expressions holds:

$$\phi_E = \frac{1}{\pi} \cos^{-1}\left(\frac{\lambda - 2}{\lambda}\right) \tag{7}$$

$$\phi_D = \frac{1}{\pi} \cos^{-1}\left(\frac{\lambda - 1}{\lambda}\right) \tag{8}$$

in which

$$\lambda = \frac{1}{2}\left(1 + \sqrt{1+\alpha^2}\right)$$

$$\alpha = b/d$$

Equations (7) and (8) are shown graphically in Fig.3. The exit pressure gradient at point C is given by:

$$G = \frac{H}{\pi d} \frac{1}{\sqrt{\lambda}}$$

This is plotted in Fig.4. The critical gradient can be shown to be

$$G_c = P_w (P_s - 1) (1 - e)$$

For noncoherent soils with space e = 40 per cent, specific gravity of particles P_s = 2.65, and P_w = 1, G_c is unity. For design purposes Khosla recommends the following factors of safety:

Types of Soil	Safety Factor
Shingle	4 to 5
Coarse Sand	5 to 6
Fine Sand	6 to 7

For greater safety, an inverted filter consisting of graded ballast and stone covered with open-jointed concrete blocks is provided downstream of the structure. The function of this talus is to prevent ersion of the chabed by action of the stream and undermining by seepage flow. For the design of this protection and also the design of the structure flow reference may be made to recent papers published by the author(3 to 13).

Khosla's method of independent variables

It is known that the effect of a sheet pile under a floor is concentrated in the immediate vicinity of the pile and dies out rapidly with distance from the pile. This means that a complex foundation may be analysed by splitting it into a number of elementary forms, considering each of these individually and then correcting for the influence of each form upon the other.

Consider the structure shown in Fig.1. In the first instance, the floor is assumed flat with negligible thickness and the two piles are regarded as independent in thier influence on uplift pressure. Under these conditions the theorectical curves of Figs.3 and 4 are applicable. For the downstream pile, calculate the ratio d/b and then find the percentage pressures ϕE and ϕD from Fig.3. For the upstream piles, calculate the ratio d_1/b and then from the same chart find the values of ϕE and ϕD. These would have been the percentage pressures had the pile been located at the downstream end of the floor, but since the pile is at the upstream end the property of reversibility of flow gives

$$\phi_{C1} = (100 - \phi_E) \text{ and } \phi_{D1} = (100 - \phi_D)$$

Thus, the lower curve in Fig.3 furnishes both ϕ_E and ϕ_C depending on the direction of flow.

The numerical values thus obtained must be corrected for (1) the effects of piles on each other, (2) the effect of the sloping glacis, and (3) the effect of finite thickness of the floor. The mutual interference of piles (Fig.5) is given by the following empirical formula:

$$C_p = \pm 19 \sqrt{\frac{D}{b}} \frac{(d+D)}{b}$$

where C_p is the correction to be applied as percentage or pressure head, and d is the net depth of pile on which the effect of pile D is required. The correction is additive for upstream points and subtractive for downstream points in the direction of flow.

The correction to be applied due to the sloping floor is

$$C_s = F\left(\frac{b_s}{b}\right)$$

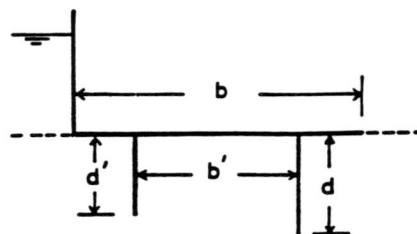

Fig. 5   *Mutual interference of piles*

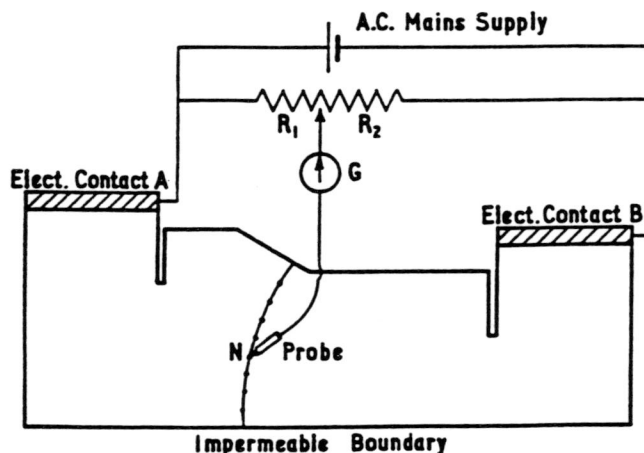

Fig. 6   *Circuit diagram for the determination of pressure lines*

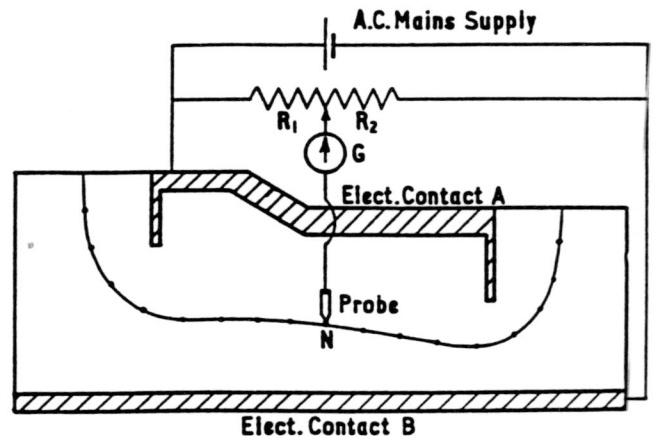

Fig. 7   *Circuit diagram for the determination of stream lines*

being positive for the down and negative for the up slopes following the direction of flow. the values of the coefficient F for different slopes (horiz/vert.) are given below.

Slope	1.0	1.5	2.0	3.0	4.0	5.0
Coefficient	11.2	8.5	6.5	4.6	3.4	2.8

In order to allow for the actual thickness of the floor the pressure between the top and bottom of a pile is assumed to vary lineary, so that the correction at point C_1 is

$$C_t = + (\phi\ D_1 - \phi\ C_1)\ \frac{t_1}{d_1} \qquad (13)$$

and at point E is

$$C_t = - (\phi\ E - \phi\ D)\ \frac{t_2}{d} \qquad (14)$$

where t_1 and t_2 are the floor thickness at points C_1 and E respectively.

Using the above procedure, the uplift pressures under the drop structure shown in Fig.1 and a regulator with a flat floor of the same dimensions are calculated and plotted linearly in Figs.9 and 10: the corresponding exit gradient G is 0.20. The required thickness of the concrete floor at any section is found by equating the submerged weight per unit length to the local uplift pressure which gives t = h/1.4. A factor of safety against uplift for the whole structure may be calulcated as follows:-

$$F.S. =$$
$$\frac{Total\ submerged\ weight\ of\ structure}{Total\ uplift\ pressure\ force}$$

$$F.S. = \frac{W'}{pg \int hdx} = \frac{100\ W'}{pgbH \int \phi\ d(x/b)}$$

The uplift force is obtained by integrating the pressure diagram graphically.

Seepage Tank

The pressure distribution and the flow lines under 1/48 scale models of the drop structure and regulator were determined in a glass-sided seepage tank 5ft. long by 18 inches deep by 4.75 inches wide. The models, each 18 inches long, were made of brass with 1/16 inches pressure tappings spaced at 1 inch intervals along the centre line. The sand, size 0.6 mm., was placed in small quantities under water and compacted consistently to remove air bubbles. Dye crystals, placed at various points in the sand, were used to trace the flow lines. The upstream and downstream water levels were kept constant by means of adjustable overflow pipes. The pressures were measured by precision mulititube manometers designed by the author

and reading to within 0.1 mm. by using a cursor with a hair line and millimeter vernier.

Electrical Analogue

The principle of this method is that eqns (2) and (3) are analogous to those governing the flow of electrical current in conducting medium. By means of a Wheatstone Bridge arrangement, the equipotential or stream lines can be plotted on a speical conducting paper ([14]), provided the boundary conditions are correctly represented.

The procedure is briefly as follows. The boundaries are drawn lightly on the conducting paper with pencil and where connection has to be made a strip of about 1cm wide is allowed on the outside of the drawn boundaries. The connecting wires (fuse wires are satisfactory) are stapled in position. The boundary strips are then given a coat of conducting silver paint and then allowed to dry.

Figs.6 and 7 show the circuit diagrams for the determination of the equipotential and stream lines under the drop structure. The variable resistor is adjusted in a certain ratio R_1/R_2 and by means of the contact probe connected to the galvanometer, a null point N is obtained. The ratio of the resistances AN and NB will be similar to that of R_1/R_2. In Fig.6 the points from a smooth equipotential or equipressure line, giving a corresponding drop in the total head for the set ratio R_1/R_2. The procedure is repreated for other values of the ratio R_1/R_2, and the distribution of uplift pressure on the underside of the structure is plotted. The flow lines are determined in a similar manner and the flow net thus obtained is shown in Fig.8. Designating N_f as the number of flow channels and N_p as the number of equioptentials along each channels, it can be shown that the seepage quantity per unit width is given by

$$q = \frac{\Delta n}{\Delta s}\ kH\ \frac{N_f}{N_p}$$

Discussion of results

The pressure distributions obtained by the seepage tank and the electrical analogue, together with the linear variation of Khosla's method, are plotted non-dimensionally in Figs. 9 and 10. For both experimental methods, the pressure variation follows a curve the slope of which is maximum for the flow under the downstream apron (0.3 <x/b< 0.8). In this region the flow channel directly beneath the underside of the structure converges (Fig.8), resulting in an increase in velocity and hence a rise in the rate of head loss due to internal friction.

The two experimental results are seen to be in fair agreement through a number of minor differences may be noticeable. In particular, near the upstream pile of the regulator structure the results of the seepage tank are appreciably higher than those of the analogue. This is probably due to the deviation of the analogue flow pattern from the true pattern in the hydraulic model, since the former kinematic and dynamic similarities exist only on a microscopic basis.

As to comparision with Khosla's method, up to x/b = 0.5, the pressures for both structures lie above the theoretical line, while further downstream they fall below. The maximum deviations from the predicted values are +8 per cent for point C_1 and -5 per cent for point E (Fig.1). The discrepancy may be due partly to experimental errors and partly to the effect of the limited depth of the permeable soil, as the theoretical curves of Fig.3 are strictly applicable to structures resting on an infinite field of

permeable strata. Experiments with the electrical analogue show that as the ratio of the depth of the permeable strata to the length of the structure T/b decreases the pressure at point C_1 increases. Over the range T/b = 0.2 to 1.0 there is a pressure variation of up to 10 per cent. For a theoretical discussion of this effect, the reader is referred to Pavolvsky's work reported by Leliavsky[15] and Harr[16].

The pressure variations along the piles are compared with the theoretical lines in Fig.11. The rate of pressure drop is seen to be much less than that predicted, but there is a large percentage of head loss around the tip of each pile, Fig.8. For the downstream pile, the slope of the pressure curve is 0.14 which is less but on the safe side compared with the predicted exit gradient of 0.20.

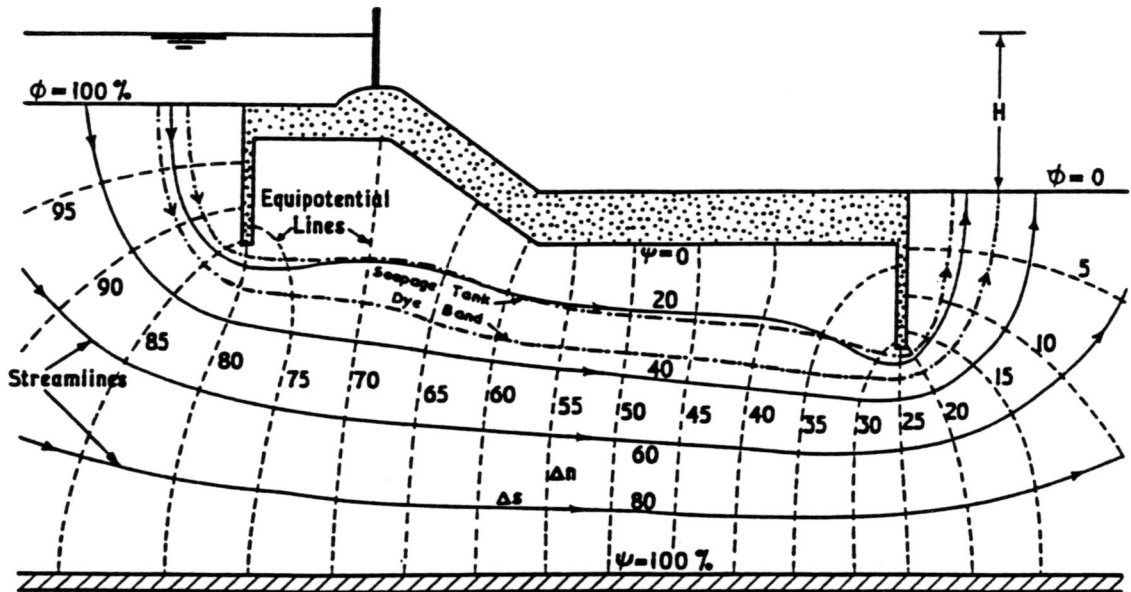

Fig. 8   Flow net under the drop structure

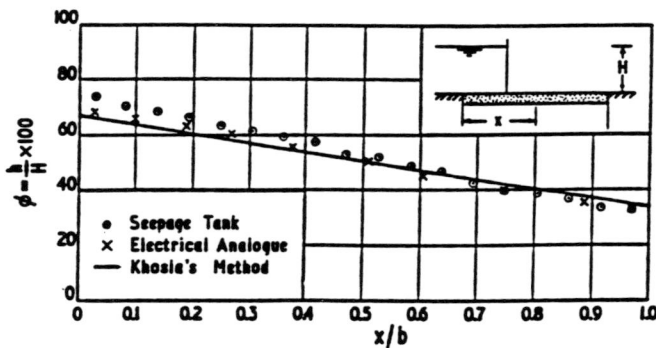

Fig. 9   Pressure distribution under the regulator structure

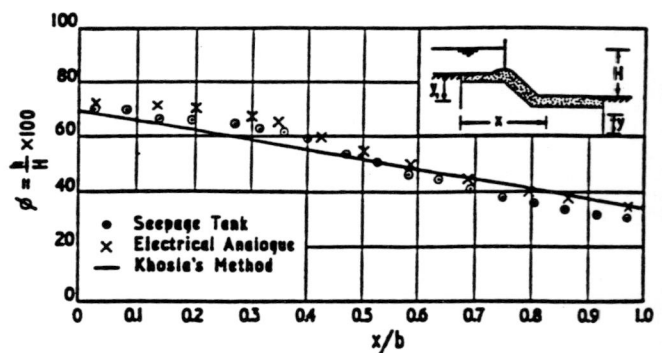

Fig. 10   Pressure distribution under the drop structure

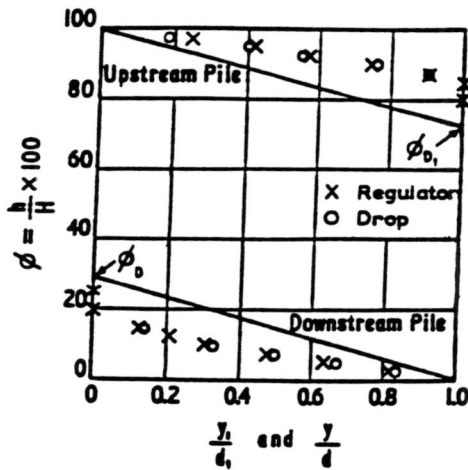

Fig. 11   Pressure variations along the piles

11.   Naib, S.K.A. : Abnormal Grouping of Large Eddies in a Submerged Jet. La Houille Blanche, No.3, p.282, 1967.

12.   Naib, S.K.A. : Spreading and Development of the Parallel Wall Jet. Aircraft Engineering (in press) 1968.

13.   Naib, S.K.A. : Stream Boundaries of Subcritical Flow in a Trapezoidal Channel Expansion. Water & Water Engineering (in press) 1968.

14.   Teledeltos Paper, Servomex Controls Ltd., Crowborough, Sussex..

15.   Leliavsky, S : Irrigation and Hydraulic Design. Vol.1, Chapman and Hall, London 1955.

16.   Harr, M.E. : Groundwater and Seepage. McGraw-Hill Book Co., N.Y. 1962.

References

1.   Khosla, A.N. et al. : Design of Weirs on Permeable Foundations. C.B.I. Publ. No.12, India, 1936.

2.   Weaver. W.: Uplift Pressures on Dams. J. of Mathematics and Physics, p.14, June 1932.

3.   Naib, S.K.A. : Equilibrium of Talus Blocks Downstream of Stilling Basins. Water Power, Vol.19, p.407, October 1967.

4.   Naib, S.K.A. : Flow Patterns in a Submerged Liquid Jet Diffusing Under Gravity. Nature, Vol.210, p.694, May 14 1966.

5.   Naib, S.K.A. : Mixing of a Subcritical stream in a Rectangular Channel Expansion. J. Inst. Wat. Eng., Vol.20, p.199, May 1966.

6.   Naib, S.K.A. : Hydraulic Design of Energy Dissipators. Water and Water Engineering, Vol.70, p.191, May 1966, and Vol.71, p.366, September 1967.

7.   Naib, S.K.A. : Photographic Method for Measuring Velocity Profiles in a Liquid Jet. The Engineer, Vol.221, p.961, June 24 1966.

8.   Naib, S.K.A. : Measuring Velocity in a Liquid Jet. Letter to the Editor, The Engineer, Vol.222, p.236, Aug.12, 1966.

9.   Naib, S.K.A. : Unsteadiness of the Circulation Pattern in a Confined Liquid Jet. Nature, Vol.212, p.753, November 12 1966.

10.   Naib, S.K.A. : Analysis of Subcritical Channel Transitions. Water and Water Engineering, Vol.71, p.55, February 1967.

PART X

HYDRAULIC OUTLET WORKS

Pipe Outlets for Irrigation and Drainage
Conduit Outfalls and
Supercritical Diffusers

PIPE OUTLETS FOR IRRIGATION AND DRAINAGE

Introduction

The US Bureau of Reclamation (USBR) developed a generalised design for stilling basins of pipe outlets [1]. It is of the impact type with the jet impinging on a vertical wall and is suitable for irrigation outlets where no tailwater is required. For high velocities there is violent turbulence and spray in the flow and cavitation damage to the baffle wall may be considerable. Francis and Gunawardana [2] carried out brief measurements on a model of a drowned tunnel outlet in which a hump was used to improve the hydraulic performance. They concluded that detailed investigations were necessary of the many variables involved.

The paper discusses the fundamental research work for the development of a new stilling basin. The aspects covered by the research include experiments on jet diffusion, flat and depressed aprons, deflection by a cill, relation between width and depth of pool and the effects of baffle blocks and the Froude number of the flow. Based on these investigations generalised procedures have been developed for a standard stilling basin design for pipe outlets.

Apparatus

The work was carried out in a special water jet equipment consisting of a 25mm diameter pipe discharging with varying inclinations, into a glass flume 1800mm long and 300mm wide. The Laser Doppler Anemometry (LDA) technique was adapted for measurements of mean velocities and turbulence characteristics, without any interference in the flow.

Vertical Jet Dispersion

The dispersion of a round jet discharging into a rectangular channel (Figs 1 and 2) has been investigated extensively by the authors and the fundamental experimental and theoretical results have been published elsewhere [3,4,5,6,7,8]. Applying these results to the practical design of a stilling basin, the indications are that the maximum rate of velocity decay is achieved when the jet impinges vertically into the channel and that 90 percent of its head is lost in a distance of 8 jet diameters.

The position of the pipe outlet relative to the back wall of the channel has little effect on the velocity decay rate but in order to obtain smooth flow conditions the jet should be placed at a distance of one jet diameter (d) downstream of the backwall.

Figure 3 shows the flow pattern for vertical jet impingement which clearly indicates a complex multiple-circulation region above the deflected jet. Observations of the free flow conditions (Fig.4) show that on impact the deflected jet spreads as a thin stream along the bed and up the side walls of the channel.

Figure 5 gives examples of the distribution of maximum velocity along the centreline of the channel for various downstream depths.

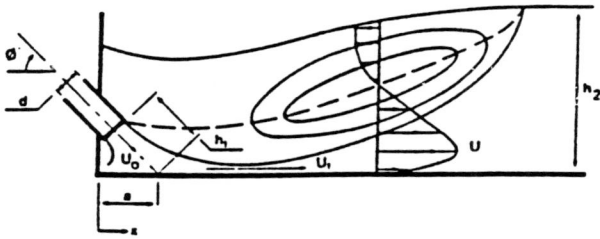

FIGURE 1

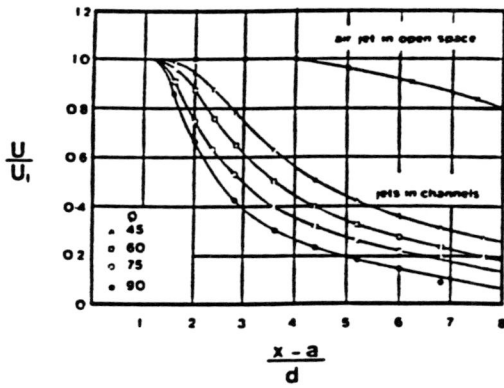

FIGURE 2

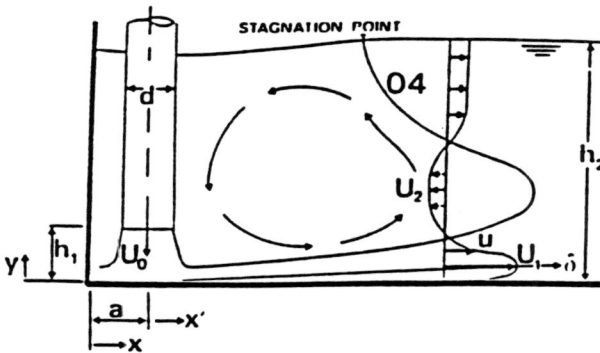

FIGURE 3

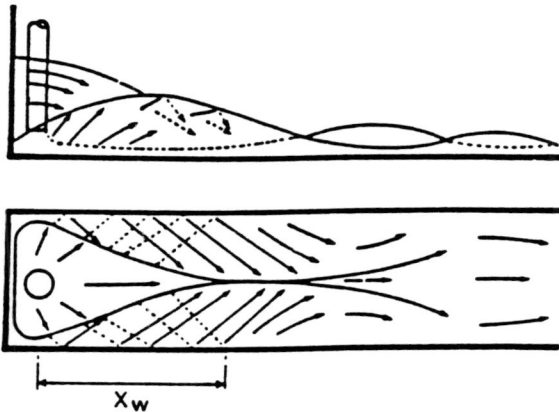

FIGURE 4

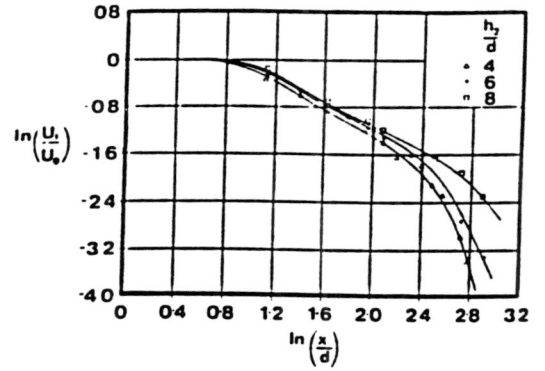

FIGURE 5

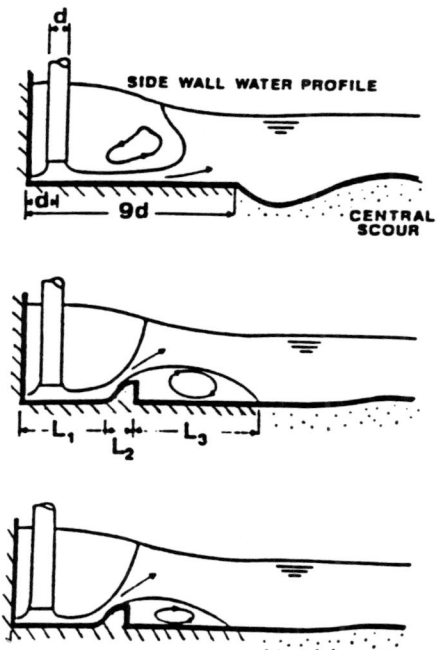

FIGURE 6

147

Experiments with a Flat Apron

A series of scour tests were carried out with a vertical jet issuing into a channel 12d wide and with a solid apron 9d long followed by a sand bed 4d deep, Fig.6. Figure 7 gives examples of the scour patterns obtained for various downstream depths. In each case the scour pattern shown is after a time period (t) of four hours. In all tests the mean diameter of the sand particles was of the order 1.5mm. Figure 8 shows the results for Froude numbers (Fo) from 4 to 8 and these clearly indicate that with increasing Fo the depth of scour increases.

In order to illustrate the relative effectiveness of the vertical pipe outlet, scour profiles produced by a horizontal and vertical jet over a time period of 5 minutes were compared (Fig.9) at a maximum Froude number of 8.

Deflection by a Cill

Square and triangular cills were tried but the most effective shape used was based on the geometry of the end cill for the United States Bureau of Reclamation stilling basin designs [1]. However, tests showed that a cill height of one jet diameter could deflect the jet efficiently in the range of depths tested.

To locate the best position of the cill, i.e. the dimension L1, a series of experiments were carried out in which the length L1 was varied between 3 and 7 pipe diameters. The least amount of scour took place at L1 = 4d.

For L1 = 4d, the length of apron required L3 to reduce scour to a practical minimum was taken from measurements of the length of the circulation region formed behind the cill. The length of this region was found to be approximately 8d for downstream depths of between 3d and 8d and a Froude number in the range of 2 to 8.

Depression of Apron Floor

To reduce the rise in water level at pipe outlet the bed of the channel was depressed to form a deeper pool. Pools depths (P) of 1d, 2d, and 3d were tested. With a pool depth of 1d results were similar to that of the flat bed with a cill. A depth of pool of 2d however, gave a considerable reduction in water level especially at low downstream depths.

The results of the tests carried out with this pool depth, channel width of 12d for various downstream depths and Froude number are plotted in Fig.10 and are compared with conditions that exist in a channel without a pool. These results clearly show the effectiveness of the pool in reducing the rise in water level near the pipe outlet.

Effect of Jet Position

An alternative method to creating a pool underneath the pipe outlet and hence reduce the rise in water level, is to move the outlet away from the back of the channel. Scour tests showed that the total length of basin would need to be in excess of 30 diameters in length. This was considered to be impractical and was not pursued further.

Effect of Basin Widths

It has been found out that the distance over which the jet travels along the bed and side walls of the channel is constant for a given rate of flow. For example, if the width of the channel is reduced by an amount r the height to which the jet rises up the side walls is increased by r/2. For practical design to compensate for the effect of a reduction in width, resulting in an increase in water level by r/2, the depth of pool should be further depressed by r/2 i.e. for Fo=8 the following width to pool depth ratios will apply:

w/d	12	8	4
p/d	2	4	6

For this design the corresponding rise in water level for each channel width therefore remains the same.

Effect of Froude Numbers

Figure 11 shows a graph of depth of pool against Froude number for widths of 4d, 8d and 12d. The data at Fo=8 is known and from earlier observations and tests at Fo=2 the flow is so calm that it is not necessary to have a pool at channel widths of 4d, 8d and 12d. Between these two Froude numbers a linear relationship is assumed to exist. For this assumption to hold the resulting rise in water level for all widths, at any given Froude number, should be the same. Intermediate points in Fig.11 were tested and the corresponding rise in water level was seen to be the same as those given in Figs. 12 and 13.

Depressing the bed of the stilling basin to form a pool obviously affects the length of apron required (L3) to eliminate scour, which was previously found to be 8d for a Froude number of 8 and channel width 12d. The length L3 was measured for varying values of w, P and Fo and is plotted in Fig.14.

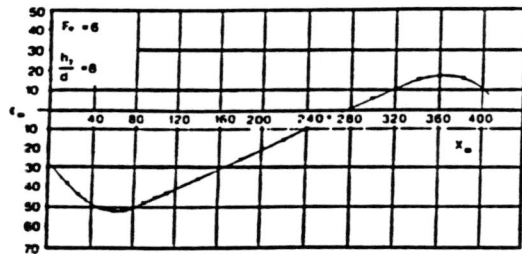

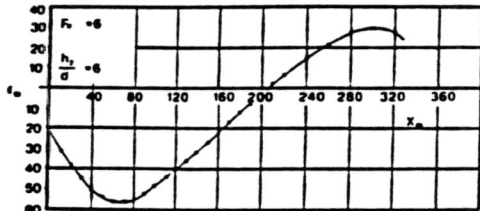

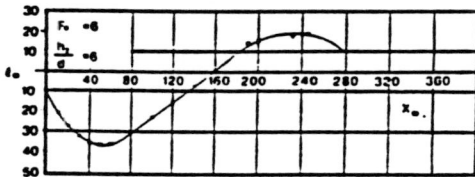

FIGURE 7

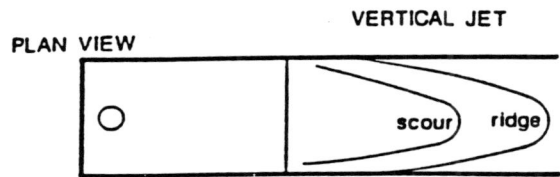

VERTICAL JET

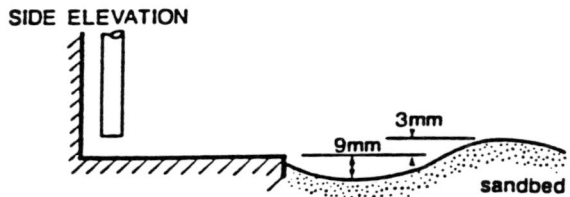

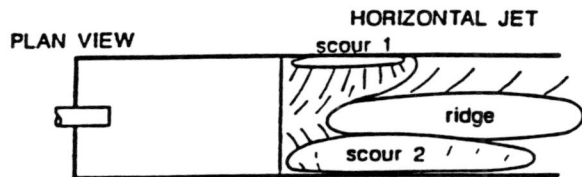

HORIZONTAL JET

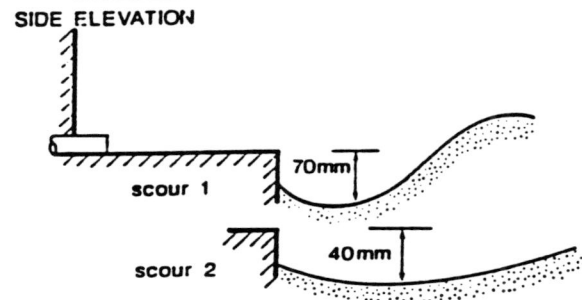

FIGURE 9

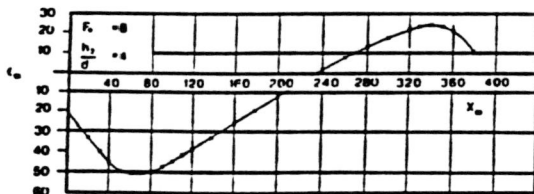

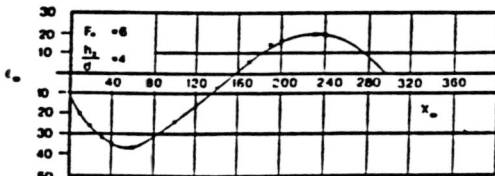

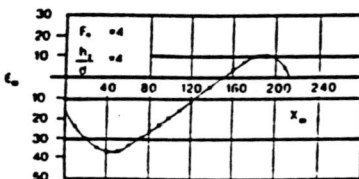

FIGURE 8

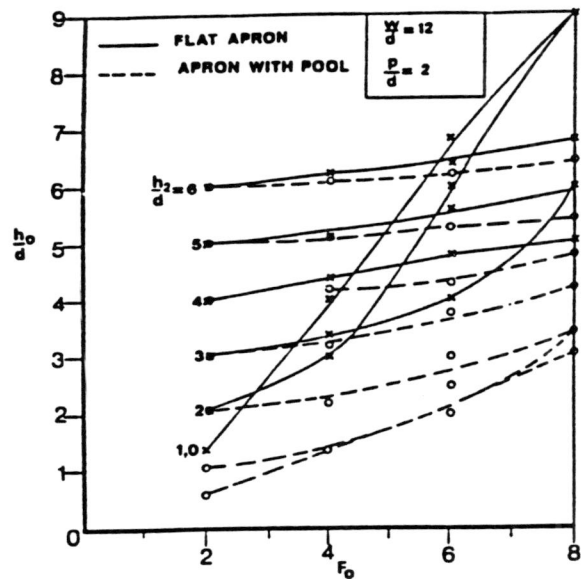

FIGURE 10

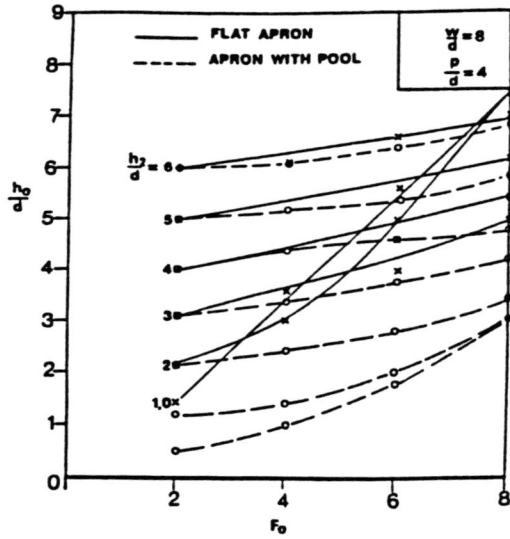

FIGURE 11

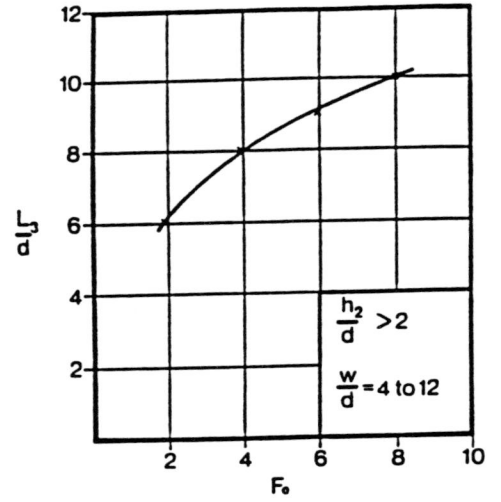

FIGURE 14

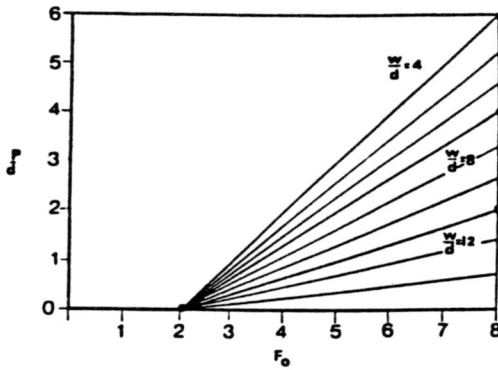

FIGURE 12

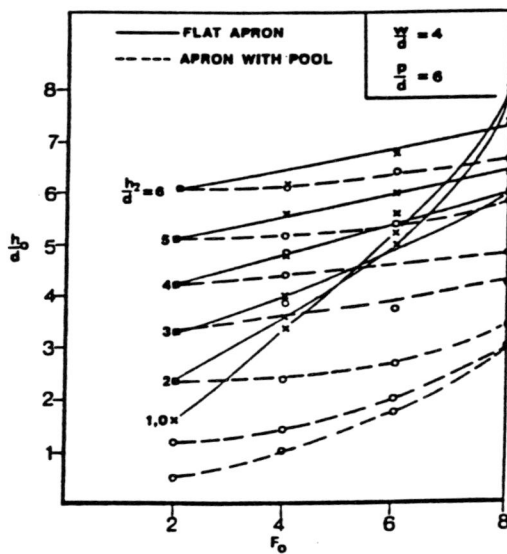

Baffle Block Dimensions

FIGURE 15

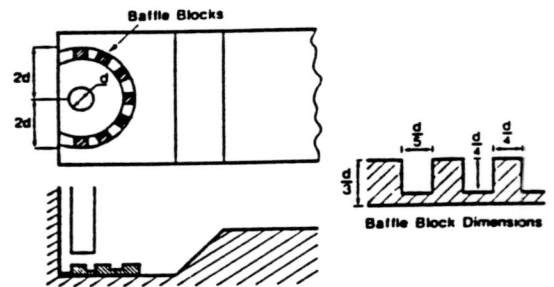

FIGURE 13

Use of Baffle Blocks

In order to achieve a greater degree of head dissipation in the pool the idea of placing a ring of baffle blocks around the pipe outlet was investigated (Fig.15). The arrangement of the baffle blocks were in accordance with the dimensions used in the Cairo Wastewater Project, i.e. the hydraulic design of stilling basins to vortex drop shafts [9].

The results of this modification was to shorten the apron length by approximately 2d. On the other hand however, the problems of maintenance that may arises due to cavitation damage to the blocks and the trappings of debris may outweigh the advantages of this arrangement. It was considered that the plain pool design would prove more cost effective for construction and maintenance purposes.

New Stilling Basin Design

Following the investigations described in the previous sections, a new stilling basin for vertical pipe discharge has been developed. It consists basically of a pool with its downstream face sloping at an angle of 45 degrees up onto a flat length of apron extending a sufficient distance to prevent scouring in the downstream channel (Fig.16).

A tailwater depth of at least 2d is required for satisfactory hydraulic performance. For the best performance the tailwater should be about 4d above the bed of the stilling basin apron.

Standard Design Procedure

To help practising engineers a standard procedure for designing new basin for high velocity pipe outlets is outlined below.

For a given design discharge Q, tailwater depth h2, and head h, the dimensions of the outlet basin are determined as follows:-

1. Compute the theoretical velocity $V = (2gh)^{0.5}$.
2. Calculate the cross-sectional area of the incoming flow $A = Q/V$.
3. The diameter of pipe is found from $d = (4A/\pi)^{0.5}$.
4. Compute the Froude number $Fo = V/(gd)^{0.5}$.
5. For this Froude number read suitable w/d and P/d ratio from the curves in Fig.11.
6. For known d, determine w and P.
7. For given Fo read length of apron L3 from Fig.14.
8. For given Fo, read height of water level ho in basin from Fig.10. Height of poolside walls = ho + freeboard (say 0.3-0.5 m). To extend 2d downstream of pool section i.e. L1 + L2.
9. Height of channel side walls hw = h2 + freeboard.
10. L1 fixed lenght 4d.
11. L2 = P.

Design Example

1. Hydraulic Data

Drainage pipe structure with a design flow Q of 5m³/s and a head h of 10m. Maximum tailwater depth h2 of 4m. Recommended width not to exceed 6d for construction reasons.

2. Pipe Diameter

Pipe diameter from $V = (2gh)^{0.5}$, therefore ($V = 2 \times 9.81 \times 10)^{0.5}$ = 14m/s, Area A = Q/V = 5/14 = 0.357 m2, d = 0.674m; select pipe diameter d = 0.7m.

3. Determination based on velocity V

(i) $Fo = V/(gd)^{0.5} = 5.4$
(ii) w/d = 6/0.7 = 8.6 For Fo = 5.4, w/d = 8.6. By interpolation between the lines for w/d in Fig.11, depth of pool P/d = 2 and therefore P = 2 x 0.7 = 1.4m.

4. For Fo = 5.4 from Fig.14, L3 = 9d therefore:
$$L3 = 9 \times 0.7 = 6.3m.$$
$$L2 = P = 1.4m$$
$$L1 \text{ (fixed at 4d)} = 2.8m$$
5. For Fo = 5.4 and maximum h2/d = 4/0.7 = 5.7. ho/d = 5.8 from curve in Fig.12. Therefore height of side walls hw = ho + freeboard = 4.06 + 0.5 = 4.56m.

The designed stilling basin is as shown in Fig.16.

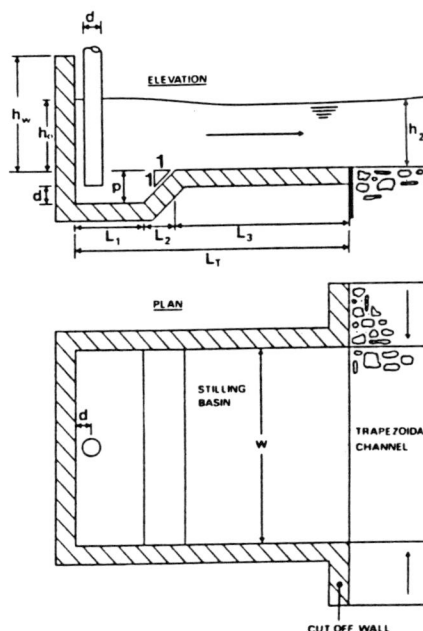

FIGURE 16

References

1.   Bradley, J N & Peterka, M; "Hydraulic Design of
      Stilling Basins : Small Basins for Pipe Outlets".
      Proc. paper 1406, ASCE, Journal of Hydraulics
      Division, Vol. 83, No. H75.

2.   Francis, J R & Gunawardana, O A : "Hydraulic
      Design of Pipe Outlets into Open Channels".
      JICE December 1969.

3.   Naib, S K A : "Spreading and Development of
      the Parallel Wall Jet". Aircraft Eng. December
      1969.

4.   Naib, S K A : "Deflection of a Submerged Round
      Jet to Increase Lateral Spreading" La Houille
      Blanche, Vol. 29, No. 26, October 1974.

5.   Naib, S K A & Sanders, J E : "Diffusion of Bluff
      Wall Jets in Finite Depth Tailwater". Discussion
      paper 18354-HY, ASCE, Journal of Hydraulics
      Division, February 1985.

6.   Naib, S K A & Sanders, J E : "Jet Dispersion
      Downstream of Pipe Outlets" BHRA International
      Conference on the Hydraulics of Pumping
      Stations, Proceedings Paper 10, Manchester,
      England, September 1985.

7.   Naib, S K A & Sanders, J E : "Jet Dispersion in
      Channels".   SECTAM   XIII   Proceedings.
      University of South Carolina, April 1986.

8.   Naib, S K A, Rasiah V, & Sanders, J E :
      "Oblique and Vertical Jet Dispersion in Confined
      Spaces". International Symposium on Jet Cutting
      Technology, Sendai, Japan, October 1988.

9.   Bramley, C E : CIRIA - Notes on Cairo
      Wastewater Project 1985. (Private
      communication).

CONDUIT OUTFALLS AND SUPERCRITICAL DIFFUSERS

Abstract

Jet dispersions in channels have been investigated in order to determine an efficient form of a stilling basin for outfalls and the minimum size of the basin for any given flow conditions. Based on these, two generalized design procedures are developed for: (i) submerged outfalls into channels, and (ii) free outfalls with supercritical diffusers. Model studies have been carried out to verify the results. The research provides practical information for engineers engaged on water projects worldwide.

Introduction

Overflow pipes and tunnels from reservoirs normally discharge into flood control channels to accomodate runoff water caused by storms and melting snow. Such channels must be provided with suitable transitions and stilling basins. Similarly, in irrigation and land drainage works stilling arrangements have to be provided for conduit outfalls and flood escapes into rivers and streams. For small waterways and cross drainage channels effective catch basins are also essential. In all these cases the same problem arises, namely the dissipation of the energy of the jet, in such a manner as to provide smooth flow with a sufficient low velocity of flow as will not cause scour in the downstream channel, or damage to the outlet structure. This paper is devoted to a brief survey of the extensive hydraulic investigations carried out by the author on jet dissipation and their applications to the practical design of submerged pipe outfalls and free tunnel outlets into channels. Model studies of an irrigation penstock outfall are presented to verify the design techniques.

Apparatus

The work was carried out in a special water jet equipment consisting of a 25 mm diameter pipe discharging with varying inclinations, into a glass flume 1800 mm long and 300 mm wide. The Laser Dopler Anemometry (LDA) technique was adapted for measurements of mean velocities and turbulence characteristics, without any interference in the flow.

Jet Dispersions in Channels

The dispersion of a round water jet discharging into a rectangular channel (Figs 1 and 2) has been investigated extensively by the author and the fundamental experimental and theoretical results have been published elsewhere [1,2,3]. Applying these results to the practical design of a stilling basin, the indications are that the maximum rate of velocity decay is achieved when the jet impinges vertically into the channel and that 90 percent of its head is lost in a distance of 8 jet diameters. The position of the pipe outlet relative to the back wall of the channel has little effect on the velocity decay rate but in order to obtain smooth flow conditions the jet should be placed at a distance of one jet diameter (d) downstream of the backwall. Figure 3 shows the flow pattern for vertical jet impingement which clearly indicates a complex multiple-circulation region above the deflected jet. Observations of the free flow conditions (Fig 4) show that on impact the deflected jet spreads as a thin stream along the bed and up the side walls of the channels. Figure 5 gives examples of the distribution of maximum velocity along the centreline of the channel for various downstream depths.

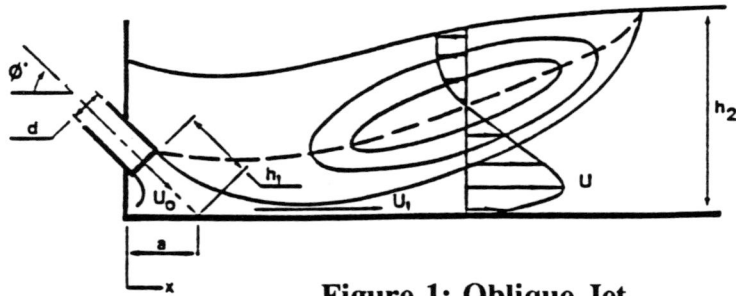

Figure 1: Oblique Jet

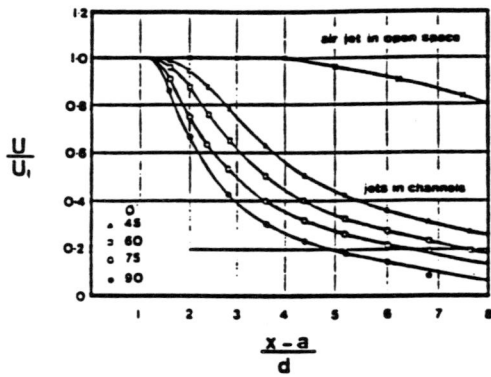

Figure 2 : Velocity Distributions

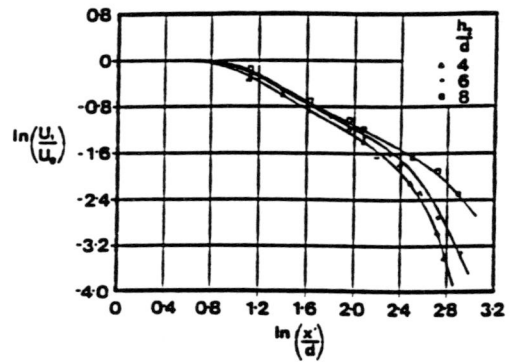

Figure 5 : Vertical Jet Decay

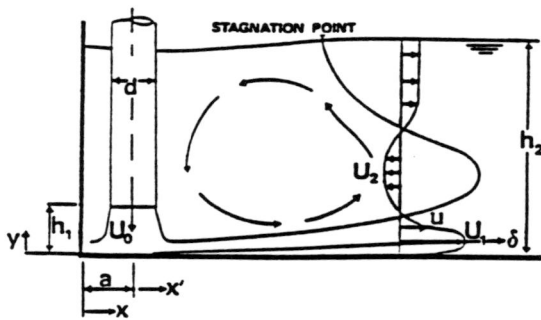

Figure 3 : Vertical Jet

Figure 4 : Free Flow

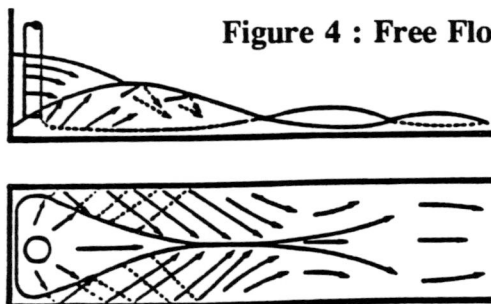

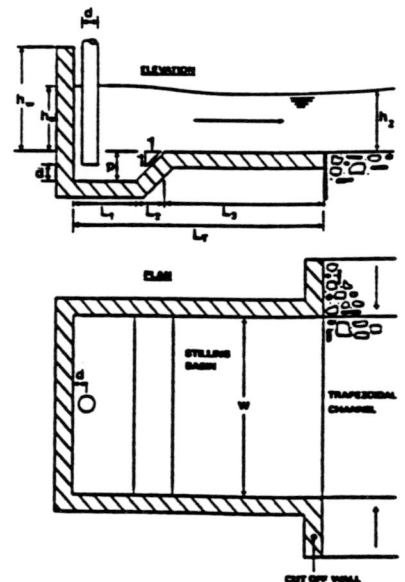

Figure 6 : New Stilling Basin

154

New Stilling Basin Development

Following the above basic research, investigations were carried out for the development of a new stilling basin (Fig 6) for vertical pipe discharge. The aspects covered included experiments on flat and depressed aprons, deflection by a cill, relation between width and depth of pool and the effects of baffle blocks and the Froude number of the flow. Based on these investigations generalized procedures have been developed for a standard stilling basin design for pipe outlets, with the aid of universal charts, (Figs 7, 8, 9). It consists basically of a pool with its downstream face sloping at an angle of 45 degrees up onto a flat length of apron extending a sufficient distance as to prevent scouring in the downstream channel. A tailwater depth of at least 2d is required for satisfactory hydraulic performance. For the best performance the tailwater should be about 4d above the bed of the stilling apron.

Design Procedure for Submerged Outfalls

To help practising engineers a standard procedure for designing new basins for high velocity pipe outlets is outlined below. For a given design discharge Q, tailwater depth h_2, and head h, the dimensions of the outlet basin are determined as follows:-

1. Compute the theoretical velocity $U_o = (2gh)^{0.5}$.
2. Calculate the cross-sectional area of the incoming flow $A = Q/U_o$.
3. The diameter of pipe is found from $d = (4A/\pi)^{0.5}$.
4. Compute the Froude number $Fo = U_o/(gd)^{0.5}$.
5. For this Froude number read suitable w/d and p/d ratio from the curves in Fig.7.
6. For known d, determine w and p.
7. For given Fo read length of apron L_3 from Fig.8.
8. For given Fo, read height of water level ho in basin from Fig.9. Height of poolside walls = ho + freeboard (say 0.3-0.5m) to extend 2d downstream of pool section i.e., $L_1 + L_2$.
9. Height of channel side walls hw = h2 + freeboard.
10. L_1 fixed length at 4d.
11. $L_2 = p$.

Design Example of Drainage Pipe Outfall

1. Hydraulic Data
    Drainage pipe structure with a design flow Q of 5 $m^{3/s}$ and a head h of 10m. Maximum tailwater depth h2 of 4m. Recommended width not to exceed 6m for construction reasons.
2. Pipe Diameter
    Pipe diameter from $U_o = (2gh)^{0.5}$, therefore $U_o = (2 \times 9.81 \times 10)^{0.5} = 14$m/s, Area $A = Q/U_o = 5/14 = 0.357m^2$, d=0.674m; select pipe diameter d=0.7m.

3. Determination based on theoretical velocity V
    (i) $Fo = U_o/(gd)^{0.5} = 5.4$
    (ii) w/d = 6/0.7 = 8.6. For Fo = 5.4, w/d = 8.6 by interpolation between the lines for w/d in Fig.7, read p/d = 2.2 and therefore p = 2.2 x 0.7 = 1.54m.
4. For Fo = 5.4 from Fig.8, $L_3 = 9d$ therefore: $L_3 = 9 \times 0.7 = 6.3$m, $L_2 = p = 1.54$m, L_1 (fixed at 4d) = 2.8m 5 For Fo = 5.4 and maximum h_2/d = 4/0.7 = 5.7. ho/d = 5.8 from curve in Fig.9. Therefore height of side walls hw = ho + freeboard = 4.06 + 0.5 = 4.56m.

Free Outfalls with Supercritical Diffusers

For large tunnel outfalls an alternative method of design has been developed (Fig.10). It consists of a short circular to square transition having a length of approximately two diameters. The shape is formed by equal spaced generators on both circle and square. This is followed by a supercritical expansion with three guide walls. Ideally, the sidewalls should be designed using the curves of Fig.11 obtained by the Method of Characterics. The angle of the equivalent straight flaring sidewalls is given by tan $\theta = 0.25/(F)^{0.5}$. The slope of the roof of the fantail section rises at a rate of 1 : 100 from the conduit to the basin to prevent air being trapped in this section.

The expanded supercritical stream enters a stilling basin with wedge-shaped sidewalls expanding with the same angle as the fantail. Moderately high velocities at the bed prevents siltation. Alternative stilling basin design of rectangular shapes with vertical sidewalls produce circulation zones on each side of the centreline and low velocities at the bottom of the basin resulting siltation.

Design Procedures for Free Outfalls

For a given design discharges Q, tunnel diameter D and tailwater depth h_2, the dimensions of the outfall are found as follows:-

1. The length of the circular to square transition is found from $L_1 = 2D$.
2. Calculate the cross sectional area of the tunnel $A = \pi D^2/4$.
3. Calculate the mean velocity in the tunnel $U_o = Q/A$.
4. Compute the Froude number $Fo = U_o/(gd)^{0.5}$.
5. For this Froude number, estimate the angle of the flaring sidewalls of the diffuser by tan $\theta = 0.25/(Fo)^{0.5}$.
    For F_1 <1, take θ = 15 degrees or 1:4 expansion.
6. Calculate the length of the diffuser, $L_2 = (W_2 - D)/2 \tan \theta$.
7. The three guide walls are spaced at D/4 and $W_2/4$ at the two ends of the diffuser.

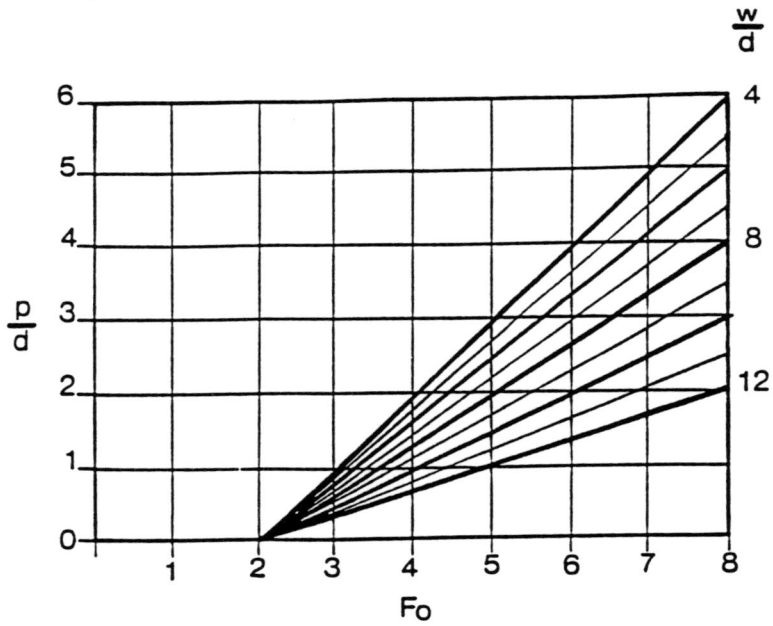

Figure 7 : Design Chart for Various Widths

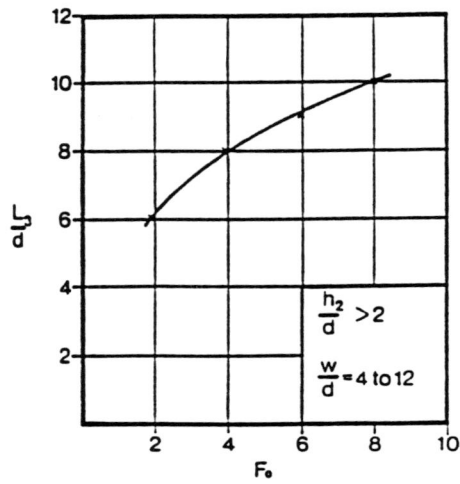

Figure 8 : Design Chart for Apron Length

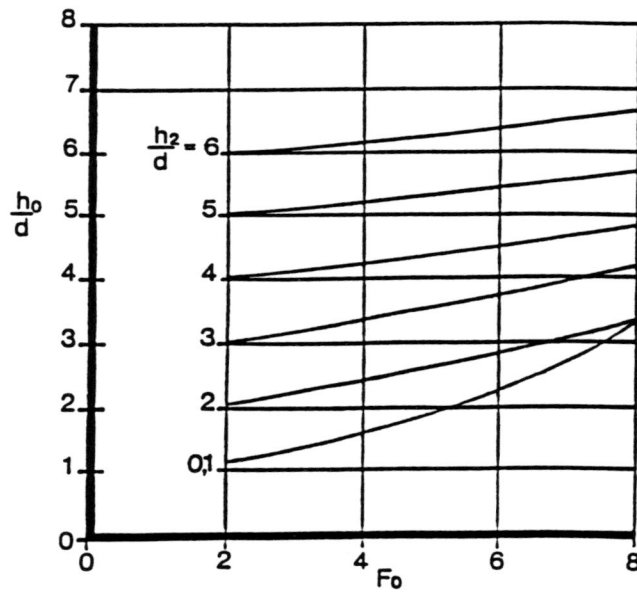

Figure 9 : Design Chart for Side Walls

FREE OUTFALLS WITH SUPERCRITICAL DIFFUSERS

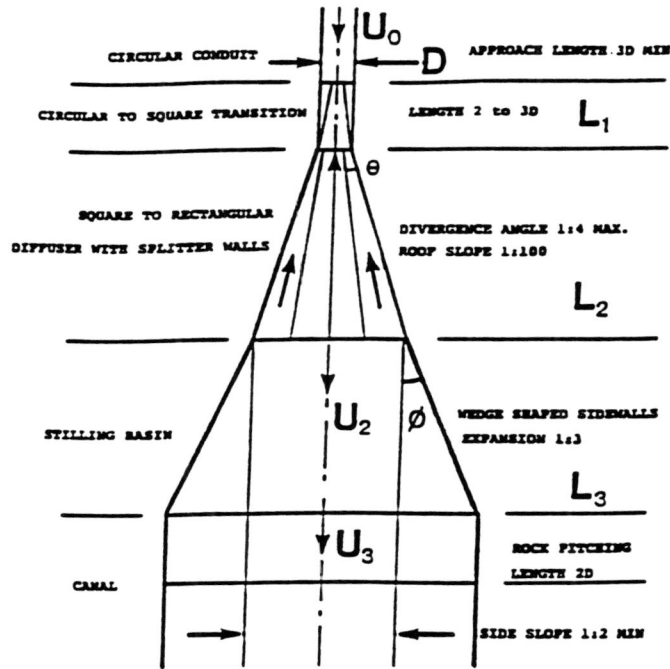

**Figure 10 : Free Outfalls with
Supercritical Diffusers**

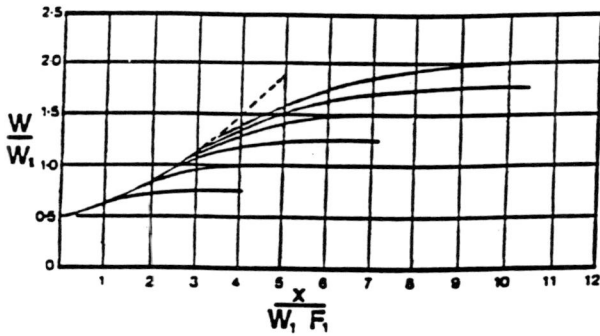

**Figure 11 : Method of Characteristics curves
for Supercritical Diffusers**

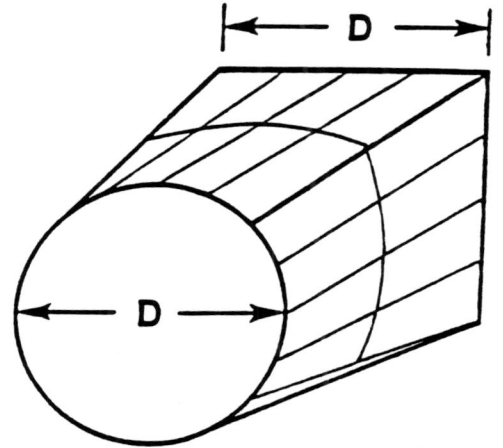

Figure 12 : Circular to Square Transition

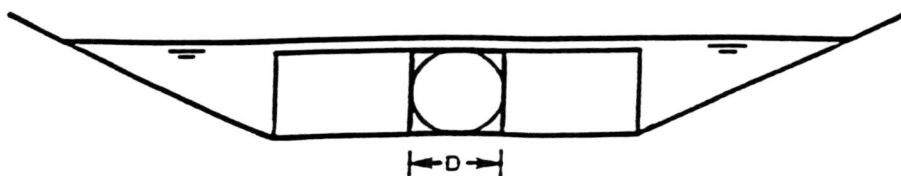

Figure 13 : Sectional View of Stilling Basin

157

8. Compute the length of the stilling basin with wedge-shaped sidewalls expanding at $\phi = 18$ to 25 degrees, $L_3 = (W_3 - W_2)/2 \tan \phi$.

9. Height of channel side walls $hw = h_2 + 0.3$ to 0.5m for freeboard.

Design Example of Irrigation Penstock Outfall

1. Hydraulic Data
   Irrigation penstock outfall 3.66m diameter with a design discharge of 50 m³/s. Downstream depth and bottom channel width not to exceed one diameter. Channel side slope to be 1:4.

2. Conduit Transition
   Length of circular to square section, $L_1 = 2D = 2$ x 3.66 = 7.32m, say 7.5m.

3. Froude Number
   Cross sectional area $A = \pi D^2/4 = \pi$ x $(3.66)^2/4$ = 10..5m², therefore velocity $U_o = 50/10.5 = 4.77$ m/s. Froude number $F_o = U_o/(gd)^{0.5} = 0.8$.

4. Diffuser
   Angle from $\tan = 0.25/(F_1)^{0.5} = 0.25/(0.8)^{0.5} = 0.258$, 15 degrees or 1 in 4 expansion of diffuser side walls. Assume $W_2 = 4D$. Length of diffuser $L_2 = 4$ x
   (4D-D)/2 = 4 x 1.5 x 3.66 = 22m. Roof slope 1 in 100.

5. Guide Walls
   Three vertical guide walls spaced at $D/4 = 3.66/4$ = 0.92 and $W_2/4 = 4$ x 3.66/4 = 3.66m at the two ends of the diffuser.

6. Stilling Basin
   Length of wedge-shaped sidewalls $L_3 = (W_3-W_2)/2 \tan 25 = (9D-4D)/2$ x 0.46 = 5.35D = 5.35 x 3.66 = 19.5m.

7. Channel Walls
   Height of channel side wall, $hw = h_2 +$ freeboard = 3.66 + 0.30 4m.

Model Studies of Irrigation Penstock Outfall

Model studies of the second method of design for a penstock conduit outfall was carried out by Taylor Woodrow International. A maximum flow of 50m³/s had to be decelerated from 3.52 m/s to one third of this velocity (1.17 m/s) over a length 30m of closed conduit and then to a mean velocity of about 0.5 m/s in the canal. The main requirements of the arrangement were that: (i) water levels in the stilling basin and canal should not overtop the banks, (ii) surface waves and turbulence in the canal should be kept minimum, (iii) head losses should be kept within acceptable limits, and (iv) siltation should be avoided or minimized.

The final design adopted is shown in Fig.14. The Froude number of the flow is about 0.60. The angle of expansion of the diffuser is 17 degrees. The angle of expansion of the stilling basin walls is 23 degrees. Head losses of

about 0.55 $U_o^2/2g$ occurred in the diffuser length.

A velocity distribution across the outfall measured by a Pitot tube is shown in Fig.16. While the velocities along the centre of the diffuser were higher than along the sides, there were no siltation in the channel. Lower velocities near the sides of the canal prevented the scouring of the side slopes. Only a small amount of turbulence was observed in the model investigations and the design proved very satisfactory.

References

1. Naib, S K A & Sanders, J E: "Jet Dispersion Downstream of Pipe Outlets" BHRA International Conference on the Hydraulics of Pumping Stations, Proceedings Paper 10, Manchester, England, September 1985.

2. Naib, S K A & Sanders, J E: "Jet Dispersion in Channels". SECTAM XIII Proceedings. University of South Carolina, April 1986.

3. Naib, S K A, Rasiah V, & Sanders J E: "Oblique and Vertical Jet Dispersion in Confined Spaces". International Symposium on Jet Cutting Technology, Sendai, Japan, October 1988.

MODEL INVESTIGATION OF IRRIGATION PENSTOCK

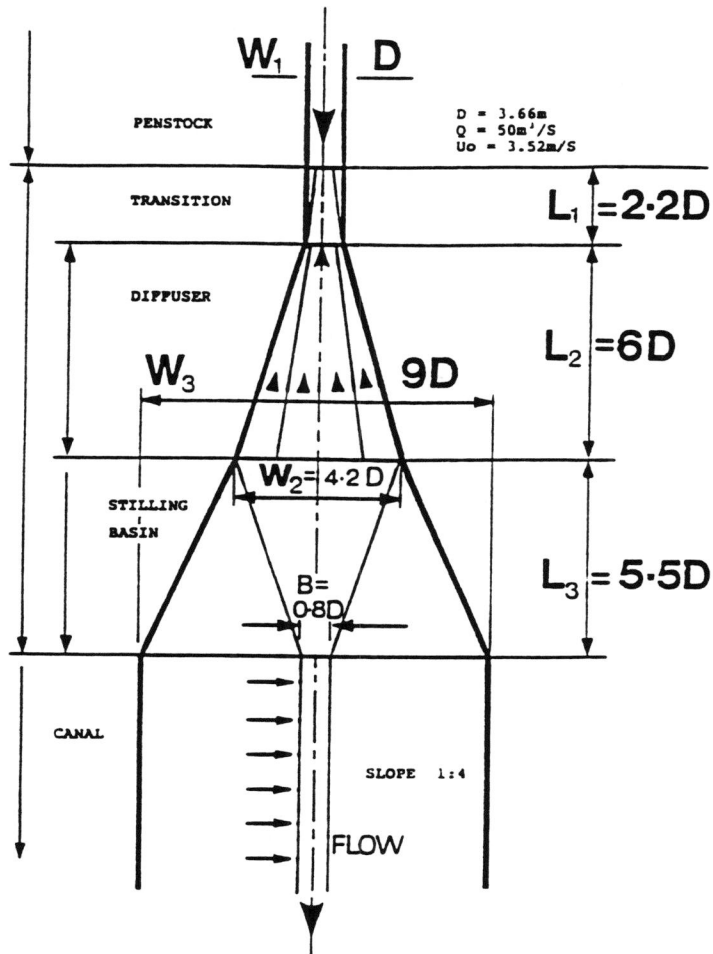

Figure 14 - Plan of Penstock Conduit Outfall

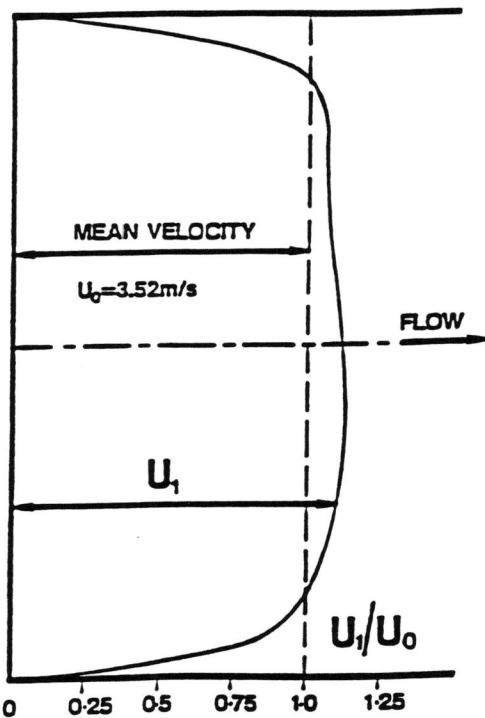

Figure 15 - Velocity Distribution Across Penstock

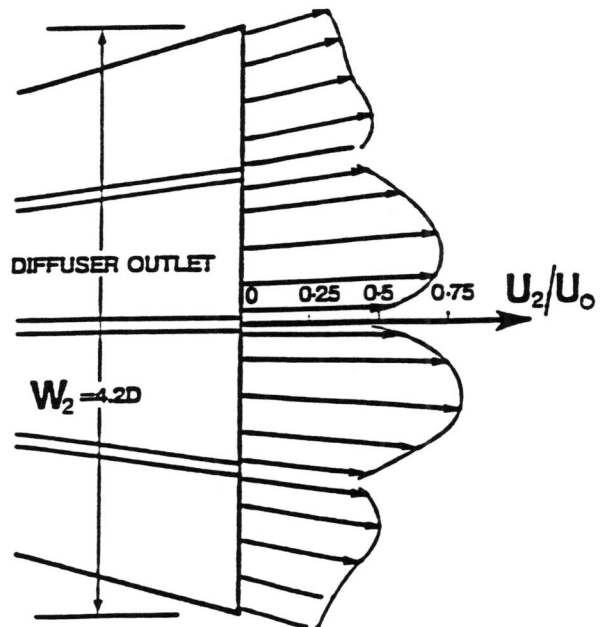

Figure 16 - Velocity Distribution Across Diffuser Outlet

159

AUTHOR'S RESEARCH PUBLICATIONS

Much of the information presented in this book is based on the author's research papers on jets and hydraulic structures listed below. Supplementary introductory chapers are provided to explain generalised theories and background design procedures, as well as to present unpublished research results useful in practice.

"Mixing of a Subcritical Stream in a Rectangular Channel Expansion". Journal of Institution of Water Engineers, Vol.20, p.199, May 1966.

"Flow Patterns in a Submerged Liquid Jet Diffusing Under Gravity". Nature, London, Vol.210, p.694, May 14, 1966.

"Photographic Method of Measuring Velocity Profiles in a Liquid Jet". The Engineer, London. Vol.221, p.961, June 24, 1966 and Vol.222, p.236, August 1966.

"Unsteadiness of the Circulation Pattern in a Confined Liquid Jet". Nature, London. Vol.212, p.753, November 12 1966.

"Investigation of Stability of 'X BOAT' Submarine". Research Report, Mirisch Films Ltd, MGM British Studios, Borehamwood, June 1967.

"Hydraulic Design of Energy Dissipators". Water and Water Engineering, Vol.70, p.191, May 1966, and Vol.71, p.336, September 1967.

"Equilibrium of Talus Blocks Downstream of Stilling Basins". Water Power, London. Vol.19, p.407, October 1967.

"Abnormal Grouping of Large Eddies in a Submerged Liquid Jet". La Houille Blanches, International Journal of Hydraulics, France, No.3, p.282, October 1967.

"Flow and Filtration Through the Wedge Wire". Research Report, British Wedge Wire Company, Warrington, November 1967.

"Distribution of Uplift Pressure Under Hydraulic Control Structures". Surveyor and Municipal Engineer, London. p.24, February 24, 1968.

"Water Heating and Requirements of Open Air Swimming Pools". Research/Design Handbook, Consultants Laboratories, London, May 1968.

"Permeability and Strength of Brixito Lightweight Concrete Bricks". Research Report, TSI Ltd, London, December 1968.

"Spreading and Development of the Parallel Wall Jet". Aircraft Engineering, London, p.30, December 1968.

"Stream Boundaries of Subcritical Flow in a Trapezoidal Channel Expansion". Water and Water Engineering, Vol.73, p.155, April 1969.

"Deflexion of a Submerged Round Jet to Increase Lateral Spreading". La Houille Blanche, International Journal of Hydraulics, France, No.6, p.455, November 1974.

"Surface Motion of a Plane Liquid Jet". La Houille Blanche, International Journal of Hydraulics, France, 1980, No.6, pp.377-383, December 1980.

"Hydraulic Research on Irrigation Canal Falls". Proceedings of International Conference on Water Resources Engineering, Southampton University, April 1984.

"Diffusion of Bluff Wall Jets in Finite Depth Tailwater". Discussion Paper, Journal of Hydraulic Engineering, Proceedings of American Society of Civil Engineers, February 1985.

"Dispersion Downstream Pipe Outlets". Proceedings of International Conference on Pumping Stations, British Hydrodynamic Research Assocation, Manchester, September 1985.

"Diffusion of the Sluice Way Jet". Proceedings of the International Conference on Hydraulic Design and Drainage, Southampton University, April 1986.

"Jet Dispersion in Channels". Proceedings of the International Conference on Theoretical and Applied Mechanics, University of South Carolina, USA, April 1986.

"Hydraulic Design of Pump Discharge Installations for Irrigation and Drainage Channels". Proceedings of International Conference on Economic Design, University of London, September 1988.

"Oblique and Vertical Jet Dispersion in Confined Spaces". Proceedings of International Symposium on Jet Cutting Technology, Sendai, Japan, October 1988.

"New Stilling Basin Design for Irrigaton and Drainage Pumping Stations". An abstract for Asian Regional Symposium on Rehabilitation and Modernization of Irrigation Schemes. The Phillippines, February 1989.

"Jet Dispersion for Overflow and Outfall Designs". Proceedings of International Conference on Drainage and Waste Management into the 1990s, Dundee, May 1989.

"Design of Pipe Outlets for Irrigation and Drainage Works". Proceedings of International Conference on Irrigation Theory and Practice, Southampton University, September 1989.

"Diffusion of Jets and Streams for Drainage Design and Management". Proceedings of the International Symposium on Stormwater Management, Kuala Lumpur, Malaysia, May 1990.

"Design of Conduit Outfalls and Flood Escapes". Abstract for International Conference on River Flood Hydraulics, Hydraulic Research Station, Wallingford, September 1990.

"Hydraulic Investigations of Conduit Outfalls". Abstract, Journal of Water Resources Management, Issue 4, December 1990.

"Conduit Outfalls and Supercritical Diffusers". Proceedings of European Conference on Advances in in Water Resources Technology, Athens Technical University, Greece, March 1991.

"Jet Dissipation in Deep Pools" Proceedings of Fourth International Conference on Laser Anemometry, Ohio Aerospace Institute and Ohio State University, USA, August 1991.

"Pressure Reductions from Pipelines to Outfall Culverts" Proceedings of International Conference on Pipeline Systems, British Hydrodynamic Research Assocation, Cranfield, March 1992.

AUTHOR'S RESEARCH PUBLICATIONS

Additional List for Second Edition 1998

"**Jets in Pipes and Chambers**" Proceedings of International Conference on Civil Engineering, Jordan Engineers Association and University of Jordan, Amman, June 1992.

"**Waterside Developments in Old Docklands**". International Conference on Port Development for the Next Millennium organised by the Kong Kong Institution of Engineers, 2-5 November 1992.

"**Masters Degrees for Construction Practice Management**". Proceedings of the World Conference on Engineering Education; Portsmouth University 20-24 September 1992.

"**The Design and Computer Analysis of Tall Structures**". Proceedings of the International Conference on Tall Buildings, Kuala Lumpur, Malaysia, July 1992.

"**Bridge Inspection and Maintenance**" Proceedings of the International Conference on Inspection, Appraisal, Repairs and Maintenance of Buildings and Structures, Jakarta, Indonesia, September 1992.

"**Dynamic Analysis and Design of Structures**" 16th Annual Energy Technology Conference Proceedings in association with The American Society of Mechanical Engineers, Houston, Texas, USA, 31 January 1993.

"**Bridge Appraisal, Maintenance and Repairs**" International Conference Proceedings on Structural Failure, Durability and Retrofitting, Singapore Concrete Institute and National University of Singapore, July 1993.

"**Diffusion of Pipe Jets**" Proceedings of the Fifth International Conference on Advances and Applications of Laser Anemometry, Dutch Association of Engineers, Veldhoven, The Netherlands, 23-27 August 1993.

"**Applications of Neural Networks in Water Resources Research**" Proceedings of the Second European Conference on Advances in Water Resources Technology and Management, Lisbon, Portugal, June 1994.

"**High Velocity Water Jets in Pipes and Tunnels** Proceedings of the 12th International Conference on Jet Cutting Technology, Rouen, France, August 1994.

"**Rehabilitation and Heightening of Dam Side Spillways**" Proceedings of International Conference on Dam Engineering, Kuala Lumpar, Malaysia, 1-2 August 1995.

"**Design of Dam Tunnels and Penstock Outlets**" Proceedings of International Conference on Dam Engineering, Kuala Lumpar, Malaysia, 1-2 August 1995.

"**Building Britain's Tallest Skyscraper**" - The Story of **London Canary Wharf Tower**" Proceedings of the Second International Conference on Multi-purpose High-rise Towers and Tall Buildings, Singapore, 30-31 July 1996.

"**Computer Programmes and Wind Tunnel Experiments on Tall Masts**" Proceedings of the Second International Conference on Multi-purpose High-rise Towers and Tall Buildings, Singapore, 30-31 July 1996.

"**Case Study and Risk Assessment of Scour Around Bridge Piers**" Proceedings of the International Conference on Role of Engineers Towards Better Environments, Alexandria University, Egypt, December 1996.

"**Oblique and Vertical Jet Dispersion in Channels**" Journal of Hydraulic Engineering, Proceedings of the American Society of Civil Engineers, Volume 123, No.5, May 1997.

Index

Authors of articles cited in the references at the end of a chapter have only been listed below if their names appear in the text of this book.

Acknowledgements and Information

Hydraulic Book Series by Professor Naib
For Education, Scholarship and Professional Advancement

Book One: Fluid Mechanics, Hydraulics and Environmental Engineering (ISBN 1 8745 36 066)

- Basic Fluid Mechanics for all branches of engineering
- User friendly concise text aimed at first and second year undergraduate and diploma students
- Fundamental principles, analysis and application to solution of engineering problems and design.
- Topics spanning the interests of aeronautical, chemical, civil, and mechanical engineers.
- Self contained chapters to allow the lecturer flexibility in organising a course.
- Material for a two year fluid mechanics and hydraulics course.
- Large number of solved examples plus problems for students to practice.
- Excellent diagrams to make the understanding of the theory and problems easy.
- Concise theory is given without unnecessary details.
- Two companion books, one deals with basic topics and the second with advanced specific topics and design.
- Advantage of continuity with the second book when teachers are recommending course books.

Book Two: Applied Hydraulics, Hydrology and Environmental Engineering (ISBN 1 8745 36 058)

- Concise treatment with solved examples at an attractive price
- User friendly handbook aimed at final year degree, diploma and masters levels.
- Comprehensive references for the practising engineer.
- Inclusion of design examples with a discussion of design principles.
- Class notes, solution manual and homework exercises for use by lecturers and course tutors
- Self contained chapters for selection by the lecturer.
- Books 1 and 2 serve the needs of students for the full period of their studies. Each book substantially stands alone.
- Helps students to gain a broad overview of the subject and also provides detailed treatment of specialist topics.
- Provides students with an authoritative reference book which they can use after they leave university.

Book Three: Jet Mechanics and Hydraulic Structures (ISBN 0 9019 87 832)

- Manual of scholarly works for research engineers, masters students and practising engineers.
- Experimental and photographic techniques explained.
- Summary of research on free turbulent jets
- Hydraulic design of control and transition structures.
- Research on rectangular and trapezoidal channel expansions.
- Research on surface plane jet, parallel wall and deflected jets.
- Research on jet dispersion in channels and dissipation in deep pools and chambers.
- Research on diffusion downstream of submerged sluice gates.
- Research and design of hydraulic energy dissipators, drop structures, conduit outfalls and supercritical diffusers.

Book Four: Experimental Fluid Mechanics and Hydraulic Modelling (ISBN 1 8745 36 090)

- Basic experiments for all branches of engineering.
- User friendly manual aimed for all years of undergraduate and postgraduate students.
- Experiments spanning the interests of aeronautical, chemical, civil and mechanical engineers.
- Standardises laboratory instructions and minimises diversity across institutions.
- The instructions may help tutorial staff to be better rewarded.
- Describes the application of hydraulic model studies for research and final year projects on a range of civil engineering problems.
- Ideal for students, teaching assistants, demonstrators and laboratory technicians.

Acknowledgements

I gratefully acknowledge the debt I owe to my institution, the University of East London, and to various individuals and organisations who have assisted in the production of this book. I am very much indebted to Professor Reginald Schofield for his kind advice and encouragement over many years. For the supply and permission to reproduce the photographs I am deeply grateful to Mr Brian Lessware of South West Water Authority and to Dr Geoff Sims of Balfour Beatty Projects & Engineering Ltd. I owe a debt of gratitude to John Jones who did the cover artwork and made valuable suggestions.

Acknowledgement must be made of considerable help I have received from my past and present research assistants. In particular I am most grateful to Dr John Sanders for the dedicated research work on jet dispersion in channels. I wish to thank Mr R Vasanthakumaran who so kindly helped in the preparation and checking of the book in all its stages. Special thanks are due to Joanna Maddison for her enthusiastic and excellent typing of the whole manuscript.

I would like to express my gratitude to Alan Hooker for his kind assistance and constant help on the operation of the wordprocessor, Ted Weedon for valuable discussions and suggestions, Phillip Plumb for general advice, Barry Nottage for library research, Simon Pattle for help with some of the illustrations, John Noble for preparing the index, Derek Merritt for his kind help and handling of the printing work, Margaret Youngman for her patience and excellent typing of some of the original papers, Anna Bass for reproductions, Sheila Johnson for administrating the research grants and Derek Hart and other technicians for invaluable help in building the research equipment.

I wish to thank firms and organisations for the useful information and photographs inserted in the book. These include UK Hydraulic Research at Wallingford, Thames Water, the former Greater London Council, BHRA, EPSRC, P & O Company Ltd., the New Civil Engineer, Mott MacDonald Group, W S Atkins, Halcrow Group, BICC Group and Cementation Construction.

Finally, I am deeply grateful to my wife, Irene Naib, for her understanding and forbearance during the compilation of the book.

Please Order through: Research Books, P O Box 82, Romford, Essex RM6 5BY, Great Britain.
